Letters from

NIGERIA

Letters from NIGERIA

ARTHUR CARLISLE

ATHENA PRESS
LONDON

First Published 2007 by
ATHENA PRESS
Queen's House, 2 Holly Road
Twickenham TW1 4EG
United Kingdom

Printed for Athena Press

I dedicate this book to my wife, Inez,
who has had to give up so much to follow me around the globe

Acknowledgements

My wife's contribution to the letters was more than advisor, chief editor, quality manager, contributor, and typist; it was her encouragement and motivation that gave me the confidence to publish these letters as a book. It became a fun thing we did together.

My thanks to the many friends in Nigeria who helped with the gathering of information and photographs, and the many subscribers who either wrote or gave verbal encouragement and made it a pleasure to compile the monthly letters.

Viva Methanol, you took me to Nigeria and paid me a salary to sit idle in an office with time to think up all the nonsense.

Preface

At the start of 2004 our lives were perfectly well organised. We had good jobs, a beautiful house which we had just finished renovating, I drove a luxury German sedan and we were surrounded by loving family and good friends. We wanted for nothing. Everything happened according to schedule; even the dogs were annoyed if routines were not followed. I felt we were stagnating and so I convinced my wife that it was time for a change.

I accepted a position in Nigeria and we resigned our jobs, sold our house and contents, got rid of the cars, gave the dogs away, said our tearful goodbyes to family and friends and left for Nigeria. There was no turning back.

I arrived in Nigeria expecting to join a dynamic team and to commence with a project to build a large chemical plant. Instead I found that there was no team, the project was on hold as the finance had not materialised, and many other promises had not been kept. The new arrivals were isolated from the rest of the company, placed in a small dingy office and given no information regarding the future of the project and my boss and I had a major personality clash. I was consumed with doubts and to further add to my woes the shipment with my personal goods had not arrived and I was living like a camper without proper kitchen utensils or linen.

I was stressed and depressed, my health was beginning to fail and soon I realised this was a downward spiral and it was up to me to turn my life around.

One hot day while stuck in the Lagos traffic amongst all the squalor, human suffering, animal cruelty and dirt, I noticed that the Nigerians around me were happy, with lots to laugh about. I decided it was time to restore my sense of humour and so I started to write these letters about Nigeria to my family and friends with the objective to both inform and entertain. Enjoy.

Letter from Nigeria One

New Home

March 2004

I have now been in Nigeria for two weeks, but it feels a lot longer, as much has happened.

I am settled into a very nice apartment, and although I have most of the furniture, I only have the bare essentials in terms of cutlery, crockery, linen, etc. Hope my boxes arrive soon. See photos attached: one of the building and the other of the swimming pool and hotel area. The hotel and apartments are one complex within a much larger complex, much like Four Ways Gardens, just much bigger, with most houses comprising two or three apartments (type of townhouses). The bigger complex has parks, shops, etc. and it is safe to walk around.

My address is as follows:

C-1 Median Apartments
Victoria Gardens City
Lekki-Epe Expressway
Ajah, LAGOS

I believe it is futile to try to post anything.

It is quite safe to walk around Lagos in the day, but the beggars pester you continuously. Driving is something quite different, with continuous traffic jams, hooter blowing, potholes, motorbike taxis weaving in-between cars, and obnoxious minivan taxi drivers. There are very few through roads, which compounds the problem.

Prices are very difficult to understand. Some examples:

- 1.5 l Coke at the supermarket is US $6.

- Case of Castle (popular local South African [SA] beer) at the local outdoor market is US $12.

- DVDs at the outdoor market from as little as US $8.

- Cutlery set (cheap), 20 pieces at the department store, US $70.

Most electronic goods are cheaper (few import duties).

This weekend is a local election and a public holiday and no movement is allowed until 16.00.

There are two other SA guys, both older but very pleasant and helpful; they stay in the same complex as I do. One is to have his family join him in July.

We have two cars with drivers between us. In due course, we will each have our own driver, with 4x4 2.4i Nissans.

I will probably be going to Singapore in a week or two.

Have not had much chance to look around yet, but will do so in future.

Hope all are well, and now you can all e-mail me.

Cheers,

Arthur

Letter from Nigeria Two

Document

April 2004

Thanks to all those who have written to me.

As this is a newsletter, I thought it would be a good idea for me to describe the adventure I had trying to arrange for my personal goods to be shipped into Nigeria. It will give a fair idea of what Lagos is like.

It all started one Monday morning when Keith (the shipping agent in South Africa) phoned me to advise that before my goods could leave South Africa, I had to do the following:

- Obtain a Form M from any bank.

- Go to SGS (the inspection authority) and give them the 'BA' number on the form.

- E-mail the 'BA' number to Keith.

That all sounded fairly easy and I thought I would get hold of my company's clearing agent and he would have it sorted in a few hours; this was not to be.

I phoned the clearing agent, and although both of us believed we were speaking English, neither of us could understand a word the other was saying. I decided to drive to his office, about 30 km away, and have a face-to-face meeting. Still I believed all would be sorted in a few hours.

The driver assigned to me at that time had two objectives when driving, namely to get to 160 km/hr as soon as possible, and then to stay there as long as possible, irrespective of the traffic conditions. The traffic in Lagos is particularly bad, very few people obey any rules, and with this driver, whiplash and a nervous breakdown are normal for passengers.

Because of traffic, it took more than an hour to get to the clearing agent. The longest hold-up was on the highway, which was not really surprising, as at one junction, two of the four lanes are used as taxi ranks and open-air market with passengers and shoppers all over the road.

We struggled to find the clearing agent's office, as the roads did not run consecutively but rather at random, and the building name and number were around the corner on another street. The building is a ten-storey high-rise, about ten years old, but a squatter camp of an office block. No communal services operate, and if you do have electricity, it comes in from the road with its own cable; the lifts are just big holes in the wall and the hallways are dark. Some offices don't have electricity, and any piece of wood available was used to make the offices and rooms. The clearing agent later confirmed that rent prices escalate the lower the floor and the closer to the window you are.

I eventually found the clearing agent's office and, as he is the CEO of the company, he has the desk at the window. His offices do have electricity, but no air conditioning. I explained to him what I needed and after a call to Keith (on my cell), he could start, but I had to go to the bank with him.

To expedite matters, I suggested we use my car and driver for the trip – no problem, but he brought along an assistant who was big, looked stupid and served no other purpose but to carry the agent's briefcase. He followed us around for the next three days. When I got to the car, I climbed into the front seat, but the agent came to my rescue and explained that the front seat was for the driver and the assistant and he and I should sit in the back. Sitting at the back is a lot easier on the nerves but the whiplash is worse, as you don't know when to brace yourself. The drive to the bank took half an hour, although the distance was less than 2 km (the traffic was getting worse).

At the bank, four security guards were on hand to assist, and both the agent's and my car doors were opened and closed for us (the assistant had to do his own), but we had to join the queue at enquiries. When we got to the front, we were informed that the agent needed a letter from me appointing him as my clearing agent. No problem, I said, give me some paper, I have a pen, I will write it now; I was informed that it must be typed. When I asked if I could borrow a computer for two minutes and type it, she replied, 'No'.

OK, I figured an hour-and-a-half delay, as we would have to go back to agent's office, type the letter and return. But it was not to be. The computers in the agent's office were all broken, and a trip across Lagos to my office was required.

After three hours of getting whiplash and quickly typing a letter, we were back at the bank, with security still giving us the royal treatment. If Nigeria returned to being a normal country and one did not need security and drivers, etc., half the country would be out of work.

In the bank, we did not get the Form M with a 'BA' number, as we did not have an insurance form. I did explain that I did have insurance that was purchased in South Africa and I had the invoice to prove it. I also went to great lengths to explain that you may not insure items more than once, otherwise it will become profitable to lose them, but it did not help. The lady informed me that the procedure says you must have a separate form and without it she, could not help me. This was day two of giving up smoking, and I was being tested.

After forty-five minutes in the traffic, we arrived at the agent's insurance broker, and a further fifteen minutes was spent on getting enough chairs for everyone to sit down so that business could begin. They could sell me insurance, and the cost was only US $22. I doubted whether I could claim on it, and in reality have probably only bought the form for the lady in the bank. The form had to be typed, and one of the secretaries was asked to do so. After half an hour, I noticed the secretary was no longer at the typewriter and was sitting next door reading the newspaper, so I politely enquired what the problem was and she told us she did not know how to work the typewriter. I offered to assist, but the boss finished the typing. These offices were not too bad, but there was no air conditioning, and as it was 86° Fahrenheit, it was hot and smelly.

Back in the car and back to the bank. It was almost 17.00, but we made it, however, still no Form M as their head office must first approve it. I ask if I can take it to their head office, but no, it must go through internal mail.

I drop off the agent and his assistant and head for home. Traffic is bad and we eventually get there at 19.00. We agree to meet the next day to go to the bank head office.

We got to the bank head office at 12.00 and the agent had found out who to see on the seventh floor. For Lagos, the building is quite good, and one of the three lifts works. The lift has a full-time driver, and on each floor there is a security guard controlling the number of people climbing into the lift.

We found the lady after completing three different registers, but she claimed the form had not arrived. The agent convinced her to check again and she found it, but it had been rejected. I was now at the point of begging, and started telling her that until my stuff arrived, I had no pillows or sheets to sleep on. It worked and she agreed to assist. She explained that as I was importing goods, there needed to be an invoice, and the Form M was part of the forex (foreign currency exchange) control. I explained that these were my own goods and I couldn't buy them from myself. It did not help – no invoice, no approval. The agent proposed a solution: let's modify the South African shipping agent's invoice to include the value of the insured goods. The invoice went from R9,000 to R59,000[1] and the bank lady was very helpful; she lent us the Tippex and the use of the photostat machine.

Finally I had an approved Form M complete with 'BA' number. One more problem: I couldn't take the form to SGS; it had to go through their internal mail, and it could only go the next day. The agent did some investigative work and found the messenger and convinced him to take it to SGS then.

The above all took about four hours, and nothing further could be achieved on that day, so back to the office. We agreed to meet the next day and go to SGS.

We met at 12.00 and went to SGS. They must be in the best office block in Nigeria; it is clean and everything works, but the

---

[1] R signifies Rands, the currency in South Africa and at the time it was R6 = US $1

agent was unable to get us past the front desk on the 11th floor. The form was there, but we could only find out at 16.00 if it was OK and if they would advise Johannesburg to ship the goods.

At 16.00, we found out the form was going to be rejected. If there had been an open window, I would have jumped out of it. The lady at reception got my tale of woe and went off for about ten minutes and returned to say they would allow the import process to begin, but the forms had to be resubmitted at the branch of the bank. I did not bother to find out what was wrong. My goods could sail, and the agent could sort it out.

He did this and phoned me two days later to say all was sorted, or at least I think that is what he said, as I still don't understand him very well.

Hopefully I will have my goods by the end of April. Happy Easter to all.

Cheers,

Arthur

Letter from Nigeria Three

Shopping

April 2004

In this newsletter, I am going to try addressing some of the things that are different and strange to us expat South Africans.

Communication and the use of English.

Although most Nigerians can speak a fair amount of English, it can be difficult to understand because:

- They tend to swallow their words.

- They string a number of words together and then leave out a few key syllables.

- They add a number of local words into each sentence.

- They use American-type sayings probably picked up from TV and movies.

Some examples:

- They use the word 'welcome' in almost every greeting, probably picked up from Americans. I was somewhat taken aback when, as I arrived home from work, the maid opened the door and said 'welcome' as I walked in; after all, is it not my house?

- The word 'afternoon' is not in their vocabulary. It is 'good morning' till 13.00 and then 'good evening'.

- Maintenance is also not in their vocabulary and is never done. Everything goes until it breaks then it stands, often for ever. One frequently sees a car that has broken down and has been left there (for years) to rot away.

- Drainage: For a country in the tropics with torrential downpours, it unbelievable, but when it rains, there is water everywhere.

- Plastic: They refer to plastic as rubber. It took some time to figure out what a 'rubbabuckit' was.

- They refer to a traffic jam (of which there are many) as a hold-up, which is not as intense and reflects the Nigerian way of doing things. If you are late because of the traffic, then so be it. The traffic can be very busy, with long hold-ups; we have taken two and a half hours to travel 22 km. This is largely due to the layout of Lagos; it has many bodies of water and few bridges, poor road maintenance, and there is no adherence to traffic rules.

A phrase we often hear from the police is, 'What do you have for me?' or, if it is over a weekend, 'Where's my weekend?' which is a polite way of asking for a bribe. I guess the police spend about 80% of their time having roadblocks and extorting money. Generally they leave the expats alone and the locals, especially the taxis, get most of the hassles.

To date I have not paid a bribe, and when they ask, I say the company does not allow us to carry cash or else act so stupid that they give up.

Shopping in Nigeria is a unique experience. There are 'Western'-style shops that are generally expensive (except for electronic goods, e.g., TVs), but you can get most things.

Then there are the formal markets, which are probably the best value for money, and you can get most of the things you will find in the average small Pick 'n' Pay[2] including a wealth of African art and tons of ivory. So much for the worldwide ban on ivory trading. The problem with shopping at the market is that there are hundreds of stalls and each owner is bugging you and you must be awake, as a white skin can add 100% to the price.

We don't buy meat and dairy at the market, but the fruit is cheap and organic.

---

[2] The Wal-Mart or Texaco of South Africa.

There are lots of little stores selling CDs and DVDs with all the latest movie titles. When in one of these stores, a colleague of mine remarked that they are probably all pirate copies, to which the store owner replied, 'It is not the origin you pay for but the quality' and then went on to show us a TV set up with a player so you can test drive the copy before purchasing it. I think he has a point. Most DVDs sell for US $7.

There are lots of stalls along the side of the road and you can get anything you want; that will be the subject of a future letter.

Another form of shopping is sitting in your car in a traffic jam. The vendors appear by the dozen, not all that different from SA, except when the traffic stops, they are there, including on the motorway, and the variety is substantial. The other day I sat in the traffic jam for about half an hour and moved about 5 km and during that time I made a list of all I could buy (I have lumped them into categories):

- Cellphone pay-as-you-go cards and many other accessories.

- Handkerchiefs, sunglasses.

- Picture frames, pictures.

- Kitchen utensils, including can openers, egg lifters, flasks, dish-drying racks, ice trays, dishcloths, jugs, variety of plastic dishes, electric frying pan.

- Umbrella.

- Briefcase, pens, calculator.

- Belts, ties, T-shirts, caps, hats.

- Telephones, desk lamps, electric and manual fans.

- DVDs, CDs, tapes.

- All the latest magazines, including *Vogue*, *Penthouse*, *Newsweek*, but no *Huisgenoot* (popular South African mag; the English version is called *You*).

- Padded toilet seat, towels, bathroom scales, hair clippers.

- Ear buds, cotton wool, toothbrush, perfume.

- Monopoly, Scrabble, Chess, a complete set of all games.

- Books on business management, finance, religion, dictionaries.

- Water and cold drink.

- Tools, including 30 m tape measures, socket sets, extension cords, padlocks, plugs, torches, levels, which will probably never be sold, as most buildings are skew.

- Car fresheners, car mats.

- Map book of Lagos which is outdated – I know, I bought one.

- Vegetables, all.

- Sweets and a lot of snacks, including meat pies.

Security is no problem. The chance of being mugged is probably more than Johannesburg, but the chance of being hijacked a lot less.

A lot of people are trying to pull scams – as I said, at the market the price can vary, and change is seldom given unless asked for and is always short, as they don't have small change. The best scam I have heard of is that during the Cold War, the Nigerian government had a secret space programme with the Russians and some Nigerian astronauts were sent into space. The programme ended before they could return to Earth and they are still stranded in space, so please send money to the following bank account, as a fund has been established to raise funds to return them to Earth.

My goods from SA have arrived in Nigeria, and I must get them out of customs, which is proving to be difficult and frustrating. Tomorrow the clearing agent and I are going to have a little talk about doing things quickly. Remember the subject of the last letter? Well, as it turns out, that paperwork is no longer required. When I have my computer at home, I will have more time to compile these letters and will include photographs of Nigeria.

Attached is a picture of the beach we go to; we rent the lapa.[3]

Cheers,

Arthur

---

[3] A lapa is a pagoda with a thatched roof.

Letter from Nigeria Four

Customs

May 2004

Since my last letter, my goods and Inez[4] have turned up in Nigeria, and as Castle (beer)[5] is freely available and cheaper than it is in South Africa, it's only the family and friends I need and my life could be complete.

In the last letter I promised to send photographs but as I have not taken all the photos and Inez has arrived and has far better typing skills and writing talents than I, I have decided to recount the adventure getting my personal effects delivered.

Cast your mind back to Letter Two, and remember that to get my stuff on a ship in Durban, I had to have a Form M from a bank in Nigeria with a 'BA' number so that SGS could telex South Africa and the goods could be sent. I met the deadline required and the shipper sent the goods to Durban.

Being proactive, I decided to avoid a rush to get things in place, so that once my stuff was here in Nigeria, there would be no delays. I enquired of the shipping agent, 'What can be put in place?' He informed me that he could take no action until he had the original Bill of Lading from South Africa. I contacted my shipping agent in South Africa, who informed me that he would not have that until the ship had sailed from Durban.

Having resigned myself to the fact that my goods would not arrive for a few weeks, I carried on with my meagre lifestyle here in Nigeria. After two weeks, I contacted the shipping agent in

---

[4] Inez, my wife, arrived in Nigeria three months after me.
[5] Castle lager is a popular South African beer.

South Africa to enquire about the Bill of Lading. We then discovered that my stuff had not sailed on the promised date. The ship was overloaded and my container was taken off to catch the next ship. Obviously the ship's captain was not Nigerian, as here overloading is not a concept that is comprehended. Of thousands of containers on the ship, why mine?

Finally my stuff made it on board, the ship sailed, and to our amazement, arrived in Nigeria early. However, any caught-up time was lost, as the original Bill of Lading sent via FedEx had not yet arrived in Nigeria. Thank God I had not FedExed all my stuff.

When the Bill of Lading arrived, off I went to the clearing agent's office (same place and trip as in Letter Two; only now the driver behaves himself, as I have hired him and he is my own personal driver).

I managed to contain my disappointment and frustration with the clearing agent, although he was extremely happy to see me, told me to take a seat and gave me a Coke, he merely put the Bill of Lading into a file. I don't know if I was naïve to think that with the Bill of Lading we would promptly go off to the port, collect my stuff, go to customs, confirm that it's personal effects, and take the stuff home. Up until meeting with the agent, my biggest concern was whether we would be able to hire a truck to move all the stuff. I was told that the Bill of Lading did not have a CRI number on it, which SGS in South Africa had to issue. I was assured that the matter would be actioned as fast as possible once the CRI number had been obtained. I contacted the shipping agent in South Africa and SGS in Nigeria and eventually it was ruled that no CRI number was required, hence no fucking Form M with 'BA' number was required either.

A few days later the clearing agent telephoned me and asked me to go with him to the port to visit the shipping company. Two hours later, he and I were at the shipping company's offices and after we were sent to about six different offices and eventually arrived back at the first one, a clerk informed me that the container had arrived and would be unpacked the next day. I asked if it would be in order if I came to fetch the stuff the next day, and he confirmed that I could come and view it. To avoid any further disappointment, I confirmed that when I see it, I take it. Imagine my disgust when I was told to take a seat and told that a long procedure had to be followed before I could walk off with my stuff. The next part of the procedure was to get the customs forms to the warehouse. One day for this, one day for customs clearance – by Friday, I should have my stuff. I had now been in Nigeria for almost two months. I was learning fast, so I decided NOT TO CANCEL my arrangements for the weekend so that I could unpack the boxes.

Two days later I had still not heard from the clearing agent and was unable to contact him on the phone, so I went up to his offices. He was not there, and none of his staff knew where he was or when he would be back. I am not sure if they knew that this answer was the best to prevent any further questions.

On Friday, the clearing agent phoned and explained to me what the delay was. The container had been offloaded at the incorrect harbour/quay and he had to pay some money to get the customs documents transferred, but all was now resolved and I must accompany him to customs. An hour later, I was in his office, had taken a seat and had a Coke, and was told his man at the harbour would phone us as soon as the customs official was in his office. Three hours later, the clearing agent disappeared and reappeared an hour later; he had been to the harbour himself to find out what was going on. The customs official had sent a message that he had been in a meeting and was now going to the mosque and would only be back on Monday. I was not sure if I was being taken for a ride, I made sure that they knew I was unhappy, but the only response I received was to take a seat and understand that this is the Nigerian way and it is not as organised as South Africa.

On Monday morning at 11.00 we were at the customs building (they officially start work at 09.00, but apparently most staff only rock up at 10.00 and get going at 11.00). We fought our way through the security; it was a rush of people being funnelled through a single door. Remember, it was 90° Fahrenheit and humidity was over 90%, and many did not have time for a bath that weekend. We got to the official's office, and it was dark, smelly, dirty, hot and completely disorganised, worse than any movie that shows a third-world police station cell. We had to take a seat in the reception area for about an hour before he and two colleagues emerged from his office. I suspect that they were having a meal. We were second on the list, and where possible, I kept out of discussions and let the agent and his man do the talking. After a pointless discussion, we were asked to take a seat outside. There were other agents in the reception area, and they told me that since this official had been promoted, he was a lot more difficult and his price had gone up significantly. After I had waited for about half an hour, his sidekick asked for my passport and went through it in detail: first it had no visa; I corrected him. Then it had expired and I had to show him the extension; then my passport wasn't valid, as it did not have a number, which I pointed out. Finally he asked, 'Where's your residence permit?' I did not have one, as I could not hand in my passport for a permit until I had received my personal effects. Keeping my cool here was difficult; each time he asked a question, e.g., 'Where's the passport number?' and I replied, 'The front page' and he stared at the front page blankly; I showed him with my finger and he merely brushed it aside with his hand. So no residence permit, what now? Must I first get a residence permit, then come back and try and get my goods released? This time I decided to take a seat and let the agent deal with this. After numerous discussions, the senior official called me in and told me that the matter could be resolved and I must listen to his assistant. In the reception area, negotiations for the bribe began. The agent knocked them down from US $360 to US $200. That settled, I had to go and inspect the goods.

At the warehouse, after about an hour, they point out my goods to me. Three large rolls of plastic wrapping, nicely crated. My seventeen boxes had been placed into three crates, and they were pointing out the first three crates they found. Shortly thereafter, my goods were identified and I was to confirm whether it was my stuff, but as I had never seen the crates before, it was a tad difficult. The lids were cling-film plastic, so I tried to lift them open to have a look inside. A little man came charging across the warehouse, saying that I could not do that, I ignored him but realised the significance of his action a few days later. I thought I would push my luck – after all, we had paid a fortune in bribes – and asked if I could take the stuff, a request that was treated with absolute humour. I then informed them that if I had an AK47 in my possession at that time, I would be leaving with my boxes. This caused more laughter (I don't think they realised I was serious). I returned to my office at 16.30 that afternoon without my stuff.

On Wednesday, the clearing agent phoned me and said that we had to have another meeting with another senior customs official and thereafter, the goods could be inspected. At about noon on my way to the clearing agent's office, sirens and flashing lights escorting a vehicle passed me (this is often the case in Nigeria, as anyone can arrange for a police escort, provided they pay the fee). I did have a chuckle, as the escort included a police motorbike (seldom seen in Nigeria) and when a taxi was too slow to get out of the way, the motorcyclist decided to kick the taxi; all this happened at about 40 km per hour. He got a rapid lesson in Newtonian physics and momentum and the taxi was not affected at all, but he nearly kicked himself off his own bike (if he was not academically inclined, he at any rate learned a valuable life skill, i.e. don't pick a fight with someone bigger than you).

At the clearing agent's office, I was informed that there was a slight delay, as the chief customs officer had come to visit this centre and was addressing the staff and the office had temporarily shut down (it was his escort that had passed me). At 15.30, we were advised that he had completed his chat and we could go to the offices.

At the customs office, there is parking for about thirty cars and they normally try to squeeze forty into it. Most cars have a full-time driver, so to move cars around is not too much of a problem. But it still takes a good half hour to get the car out. I suggested we get off in the road, walk about 30 extra metres and when we were finished, meet the driver there and avoid the delay. But no one else would agree to this. In Nigeria, status is very important, and you don't walk from outside if you have a VIP pass to park inside.

Back into the customs office building and into a queue waiting to see another senior official. Same rush at the door, except it was now in the afternoon, 90° Fahrenheit outside and still no one had taken a bath. Finally in to see the customs official, only to be informed that it was now too late in the afternoon and I must come back tomorrow. It was at this stage that I told the customs official that my solution for Lagos is to take a big bulldozer and push it all into the sea. He seemed to take it with some good humour, explaining the Nigerian way again, but it cost us later. Same issue with getting the car out again.

The next morning, off to the customs office, I waited in the car park and the clearing agent went to collect the customs officials. Four of them arrived at my car with the clearing agent. The warehouse was about 1.5 km away, and we all had to drive in my vehicle, which happens to be a Nissan Double Cab. The status thing comes into effect again, and me, my driver, the clearing agent, his assistant and four customs officials all piled inside the double cab, four in the front and four in the back; few had taken a bath since the weekend. I offered to sit in the back, but they insisted there was enough room inside.

We got to the warehouse and waited. The customs officials disappeared and eventually I went to my crates and started opening them, and again the little man came charging across the warehouse, objecting. I then found out that he was the carpenter and this was his job and a bribe had to be negotiated before he set to work opening the crates. The customs officials arrived and wanted my boxes removed from the crates, so I start taking them out, only to be admonished and told, 'That's not your job.' Then two guys turned up; they are the labour. Again bribes were paid, and the boxes were taken out of the crates (10 minutes' work).

The customs officials had me open every box and pack everything out all on my own. Picture this: there were now about fifteen people watching me do this, either in my employ or bribed or about to be bribed, but this was not their problem. When all was repacked, the clearing agent and the customs official took a little walk and more bribes were paid. I then enquired, 'Why did I have to unpack everything if we had to pay bribes?' The answer was that the bribe would have been more if I hadn't unpacked, as it would appear that I was hiding something.

Just as I thought the final hurdle had been crossed, two ladies with incoherent accents turned up; one looked like a kitchen maid and the other one was dressed like a girl guide in a khaki uniform: the Nigerian Drug Enforcement Agency had arrived, and the whole process of unpacking the boxes had to restart from scratch. In the second box was a feather cushion, and the girl guide wanted it opened. I would have understood this a lot better if I were a Nigerian trying to take my personal goods into South Africa; after all isn't this where the drugs come from? I was now getting used to Nigeria and took one look at the clearing agent, nodded my head, and he and the Nigerian Drug Enforcement Agency took a little walk. I did ask these two ladies, 'Do I really look like a drug lord.' But their blank expressions never changed.

The boxes were repacked, and bribes/dashes were paid to the forklift driver, warehouse manager and some other soul there, purpose unknown.

Yes, you all know it, I still did not get my boxes then. We went off to see the senior customs official to whom I had explained, the previous afternoon, how to solve Lagos's problems. After an hour, the customs officials from the warehouse had completed their report, which amounted to a list of contents. If they had looked in their folder, they had a far more comprehensive list, which I had given them previously. This clown reviewed the report and then said if I wanted the stuff, he wanted payment in euros. The clearing agent made some joke about a politician recently being in trouble for receiving overseas money. He wanted US $140 and after a long negotiation, he was paid US $140.

I returned to my office and again did not cancel my weekend plans. On Friday afternoon at 17.30 (yes, I was in the pub), the clearing agent phoned me to say that he had my goods and he was at the gates of the port. We agreed we would meet when he was closer to my location.

Finally we met about 20 km from my house at 20.00. At 21.30 we arrived at the estate where I stay, having passed through three road blocks, and having to pay a bribe for the truck at each one. The security at the estate entrance would not let the truck through the gate, claiming it was after 17.00. At this stage I had had enough and put my foot down. Eventually we got through the gate and had the same fight with two other security check-points, but no money was paid. At 22.15, we got to my apartment.

The whole thing cost about US $1,700, about 25% more than

getting the goods from Middelburg to Lagos Harbour. That is why this country will stay poor.

I can assure you I have not added or embellished any part of this story, and in fact, due to the time that has passed, I have forgotten some of the events.

The good news is that there was no damage to any items, except for the glass on three pictures and one champagne glass, which was the result of our poor packing.

As I pointed out earlier, Inez has joined me and the apartment is beginning to look like home.

I should be off to Indonesia and Singapore for two weeks early in July and hopefully will complete another letter before I leave.

To all of those who have written, thank you very much, and those who have not, you will have to start paying a subscription for 'Letters from Nigeria'.

See picture attached of the truck that brought my goods home.

Cheers,

Arthur

Letter from Nigeria Five A

Life in Lagos

June 2004

A few weeks have passed since my last letter, and I promised a letter before I leave for my trip to Indonesia, Singapore and now also India.

I leave on Friday and just won't be able to put a full letter together, so here is the first half, the second to follow when I return. However, I might do Letter Six about my trip and Letter Five B later.

The picture above many have seen, and I have shown it to make all jealous. It is a nice place, and although we have an apartment, it is part of the hotel and we can use all the facilities. I swim at least four times a week, do a 5 km walk no less than twice a week, play action cricket once a month and soon will be playing tennis and squash. My weight is a bit better but not where it should be. The reason for this can be seen in the next pictures.

Yes, it has a nice pub and Castle lager is available. That's the nice part; now let's see some real Lagos.

Lots of nice dirt and squalor. Yes the top picture is of a butcher, and the one below, a typical Lagos scene.

In Lagos, we have, on average, one power failure a day, but looking at the distribution infrastructure, it is not surprising.

The telephones are not much better; there is normally only a 50% chance that an outside line will work.

Lagos has a very tropical climate with lots of rain and big storms. Although this has been going on for years, somehow no one has tried to sort out the drainage. When it rains, Lagos stops. Last week it took me over three hours to get home, and the next day I could not get to work at all.

And the next picture will make some people very happy.

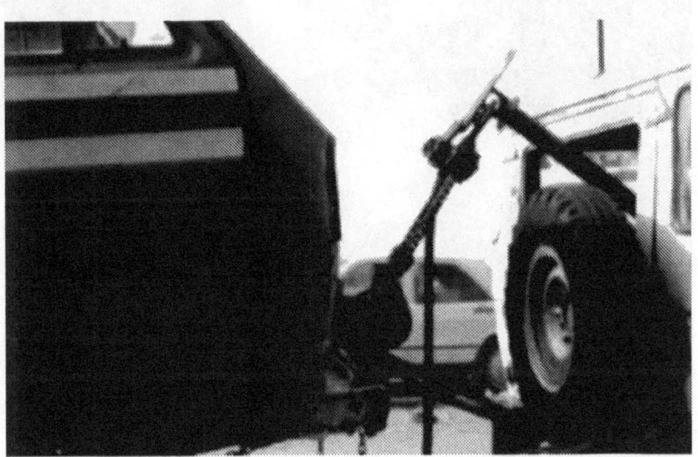

The above is the standard breakdown in Lagos. The loading is wrong, and in most breakdowns, the chassis is bent.

This car has not been in an accident, nor has it rolled; rather, it is in the shop for repairs. Who needs a lift or a workshop?

On Saturday, we were invited to the beach by some friends. To get there you take a boat ride through the harbour and up the lagoon. Our boat is being launched the Nigerian way.

Inez playing with big man's toys at the beach

Speedboat

The tropical paradise full of rubbish

This waste on the beaches is washed up from Lagos (all beaches I have seen look like this up to 70 km away) and often include medical waste. It has taken me some time to realise that Nigerians are a very wasteful society; if something breaks, dump it; rubbish, throw it out the door and never worry about the consequences. We find the same with our staff, they will waste soaps and any other consumable, especially if they are not paying. Hygiene products are generally expensive here.

We all think there is no money in Nigeria? I am not always sure if that is correct.

A church building in Lagos

Evidently there are a lot of people who know what is wrong, just not enough who care.

We hear more and more from Nigerians that it is time for a strong military dictator. I don't know if that's right, but I do know democracy is not the solution for Nigeria.

Inez's 40th birthday party, for which she had to prepare and do all the work and did not even make it into a photo. The people in the next photo are yours truly (for those of you who have forgotten) and friends in Nigeria.

As with all South African parties, a braai (barbeque).

My time is running out, and I leave for my business trip in two days' time. I will be back around 15 or 16 July. Inez will be at home (that is, in Nigeria for us).

Cheers,

Arthur

› *Letter from Nigeria Five B*

Life in Lagos

August 2004

Inez and I are doing well, and as time goes, we are settling down, and things that were strange and absurd are now becoming the norm for us. Although we won't always be shocked or surprised, our values will hopefully not change. A good example of this is corruption, and although I partake in the practice of paying bribes, it is when I have no other choice, and I will never accept a bribe.

I am sure some of you must be thinking 'no other choice' is open to a broad interpretation and is a tad convenient, so here are some examples when there is 'no other choice'.

1. NATIONAL ELECTRIC POWER AUTHORITY (NEPA)

NEPA is the Eskom of Nigeria and also known as No Electric Power Again.

One Monday morning, the security guard at the gate called me to say that the cables were burning outside.

On investigation, I found the connection of the cable supplying our building and the overhead lines had a loose joint that had become very hot and the insulation had started to burn. I issued an instruction to start the diesel generator and take the supply for the building from there.

NEPA was contacted, as the connection was on their equipment and they had to rectify it.

The next day an electrician from NEPA turned up at the office, claiming that there was nothing wrong with their equipment and fault was with us – no problem; for a small fee he would rectify it for us.

I then went to the trouble of explaining to him what the problem was and how he should rectify it. *Good free advice*, I thought, *and is he not lucky that I am here to help?*

He disagreed with my analysis of the problem and my solution and insisted his services were required. *Shame*, I thought, *he does not understand. I will have to enhance my explanation.* I took him down to the building switchboard, showed him the gauges, and with pen and paper went into a detailed explanation.

He responded by saying he understood what I was saying, but I had to remember that in Nigeria, electricity works different to the theory, and it was essential that we get him in to carry out repairs on our building. (I wanted to say to him that it is the people like him who behave funny in Nigeria and not the electricity.)

I began to realise (I thought I was a fast learner) that we were not going to get the problem fixed unless he got some private work, so I decided to make the best of it (running the diesel generator costs double that of NEPA) and I negotiated an agreement that, besides the repairs to our building, he would also repair the connections on the overhead line.

I then realised that some of our administrative staff were also involved as his fee was ready to be paid out before our negotiations were complete.

I decided that when the electrician from NEPA came to do the repairs, I was going follow him to see what he does and then prove to him that the theory of electricity is valid in Nigeria.

Two days later, when he had not returned to carry out the work, I followed up and was informed that he would come on Saturday.

Saturday I got to work early and by 13.00, no one had turned up, so I left.

On Monday I got to work and was informed that the connection had been redone and the wiring of the building fixed. I went to check, and the connection was redone, but all the gauges on the switchboard were disconnected; no more theory for the electrician.

So what did I learn from this? I am a slow learner, and if I had just kept my mouth shut and let him get his fee, he would not have had to wait until I was not there to fix the wiring (disconnect the gauges) and saved three days' diesel.

2. BUYING PETROL

Something we have never experienced in South Africa is a fuel shortage. Even in the times of big strikes in the early 90s it was never a problem. In Nigeria, if there is a general strike, very few people go to work, there is no public transport, and most shops stay closed. Before such action, those who can afford it start hoarding food and fuel.

Thus, just before a strike, the petrol stations experience a rush of customers and long queues develop. The staff at petrol stations, instead of working harder, slow things down until the queues are very long and then start a queue-jumping service, and the more you pay, the faster you get served. No pay, no service.

Off the topic, but it is interesting to note that the 'no pay, no work' rule does not always apply here, and during a general country or industry strike, employees (of government and large companies) will get paid, as it is deemed the strike is not the employees' fault, and the fact that they did not work is irrelevant. Most people scrape out a living by working in the informal sector, trading, farming or supplying a service (e.g., cobbling), and it is these folk who are hit the hardest, and after a few days, they start to starve, as they have no food and no money.

The problem with corruption here is that Nigerians accept it and it is part of their culture. They don't mind paying bribes and view those in a position to extract a bribe as 'lucky' and hope that one day things will change and they will be in a position to extract bribes.

At roadblocks, the police generally leave us expats alone, and it is the taxis and trucks that are most affected. I have pointed out to the locals that it is not the drivers who are paying but ultimately the commuters or consumers. They, on the other hand, say it is better that the police extract bribes to supplement their income, as they are paid badly and sometimes not at all, and if they don't earn money in this manner, they carry out armed robberies, etc.

In Nigeria, there is no point in making laws or rules. Even if it is for the good of the country, it merely results in someone being made to extract another bribe.

We live in the Hyde Park of Lagos (for my European/American subscribers, that is the Beverly Hills of Lagos), and a substantial number of new houses are being built in our neighbourhood. Inez is still an active walker, and for some reason seems to enjoy it more when I accompany her, suffering from the strenuous exercise. However, it has given me the opportunity to view the architecture and building techniques of Nigeria, and I have come to the following conclusions:

- Although spirit levels are easily available at the shops and at the market, they are never purchased.

- Squares are adjustable anywhere between 70–110 degrees, but never at 90 degrees.

- Architects only get taught one style of house at university.

- If drawings are made before the building commences, they will probably bear little or no resemblance to the final structure.

- Symmetry is only achieved by coincidence.

- Quality is not important.

- Concrete slabs have more cavities than Swiss cheese.

Below are some photos to demonstrate.

Besides the awful colour, every arch has a window centralised on the inside. I am not sure if behind this wall is a toilet or if they didn't have enough windows or just plain forgot, but for some reason, this particular arch did not get a window. Also, look at the columns, and you will see the one on the right is not quite in the right place.

Look at the balconies on the front of the house, and it's not an illusion of the photograph; they do slope, including the roofs. It almost looks as though they are loose and are about to fall out. Look at the eves on the side of the house; they could not have got them more cockeyed.

Gates are big business and something of a status symbol (in SA, it is a swimming pool and lapa). In my view, the gates in this next photograph are the epitome of bad taste.

To all the subscribers who have paid their subscriptions, thank you very much. Some have been very good, while others have large outstanding balances.

The weather has been excellent for the last two months, it's between 77° and 82° Fahrenheit, partly cloudy with occasional rain. Humidity has been reasonable, and it is strange, but I have not yet experienced a winter this year.

In my next letter I'll deal with transport in Lagos.

Keep the e-mails coming. It is nice to stay in touch.

Cheers,

Arthur

Letter from Nigeria Six

Travels in the East

July 2004

It has been almost a month since I last wrote and then only half a letter. I did promise to complete Letter Five A with Five B and send it soon after I returned from my trip to the Far East. Unfortunately I am not going to post Five B due to the following:

- I have not taken all the photos.

- The format of sending the letter and ten other mails with the photos is clumsy.

- I don't feel guilty, as most subscribers have not paid their subscriptions by writing to us.

- Letter Six will be sent now and Five B in a week or two.

- In twenty-one days I was in four countries, caught eleven flights, lost my luggage twice, and spent a fortune on goodies for Inez and me.

I was due to fly out from Lagos on Friday evening, and as the traffic is unpredictable, I left early. On this particular day, there was no traffic and I was at the airport very early. I checked in, went through immigration, and went to get a bite to eat. The first restaurant referred me to the other restaurant upstairs, as they no longer served food.

At the upstairs restaurant, I was informed that I could have a drink, but they couldn't serve me, as they had no food (i.e., not certain selected items on the menu are not available; they have no food). When I asked why and when would there be food, the reply I got was, 'It has not arrived and they do not know when it will.'

So I settled for a bottle of sandwiches (as many of you know, I don't drink beer voluntarily; I am normally forced).

About ten minutes later the six restaurant employees sat themselves down at a table and had a hearty meal. I decided it did not help to get cross, and I should rather try to help. I called over the head waiter and asked him to spot the mistake; needless to say, he couldn't see anything wrong. I explained that there was no food for the two patrons in the restaurant, but the six staff members were all eating, and after all, if a restaurant can't serve food to its customers, it has no reason to exist. He, on the other hand, had a very different angle on it: the staff were hungry and the food came for them, so they ate, and it was unfortunate that the customers couldn't buy food, but at least the staff were not hungry. I gave up and had another bottle of sandwiches.

This story has a happy ending, as nine hours later I was in Dubai, and was able to gorge myself on my first Big Mac since early March.

Next stop was Singapore, and then on to Indonesia. It was my first time in Indonesia, and I was pleasantly surprised. I can recommend it as a destination for anyone's next holiday abroad. The reasons are as follows:

- It is very cheap; you can stay in a four-star hotel for less than US $50 per night, fly between cities on a good airline for less than US $50, or hire a car with fuel and driver for US $35 per day.

- The infrastructure is good.

- The people are relaxed.

- The large shopping malls are cheap and your wife can't buy much, as the Indonesians are tiny and we Westerners are just too big for the clothes.

- It is exotic, different, and there is lots to see.

- Great weather and beaches.

For those of you who know Barbara and Carol,[6] there are lots of their type of 'local' restaurants. On the other hand, there are few fast-food outlets, and it's the only place in the world where you can buy rice and chicken at McDonald's (they don't eat bread only rice).

Indonesia has the world's fourth-largest population after China, India and USA, and although it is busy, you don't feel overcrowded.

South Africa should start playing rugby against Indonesia, and because of the size difference, we are bound to win and can still claim to have beaten a South Pacific country.

The company I visited went out of their way to be hospitable, and every night was late.

Next stop was Singapore, and as usual, everything was super-clean, over-organised and somewhat expensive, especially the beer.

My hotel was near the Singapore River and hence near all the entertainment areas, which was convenient, but not so good for the waistline. I ended up watching the British Grand Prix at Harry's Bar.

I don't know if you are aware that Singapore has some very high mountain ranges; I wasn't until Inez's cousin Annabelle, who lives in Singapore, invited me to a walk. As it turned out, it was a marathon cross-country event, and we crossed the mountain range a few times. I nearly died. Afterwards there was an excellent function at which I could restore all the fluid lost during the marathon.

Thanks to Annabelle; she made my stay very pleasant and enlightening.

I am not the greatest fan of Singapore; I find the people brainwashed and bland. Everyone tells you how safe and honest it is, but everything is double-locked and I was still ripped off by half the taxi drivers and had airport staff trying to get me to carry contraband for them. You can find ivory to buy in the shops.

---

[6] Barbara is my sister and Carol, a good friend, with whom we have travelled abroad. They always try to take us to the most 'local' restaurant, e.g., the canteen and the Kuala Lumpur bus station.

Singapore Airport reminded me of my first few days in the Army: if you don't know where to go, just do something, anything, and half a dozen people will come running to shit on you and tell you what to do or where to go.

I wanted to buy a book and struggled to find a bookshop, and when I did, the general quality level was poor.

Next stop was India, and it was worse than I expected. The ordeal started on the flight to India. The flight was packed with migrant workers; each had bought two huge bottles of Chives Regal at the duty-free (I am sure it has a good resale value in India), and there was just not enough space in the overhead lockers, so hand luggage was all over.

They served a meal on board (you had a choice: curry with or without mutton), and as the steward pushed the trolley down the aisle, hands would reach out and nick the food off the trolley. Later, when I was waiting for the baggage, I noticed a lot of duty-free bags stuffed with aircraft meals and drinks.

India has a reputation for bureaucracy, and I was warned that when you get off the plane, hurry up and get to immigration first or stand in a long queue (hours). I was not the only one aware of this. Some tried too hard, and as the plane's wheels touched the ground, about half the plane jumped up to open the overhead locker before the plane applied brakes, and when it did, the whole lot came flying down the passage, the Englishman sitting next to me was in hysterics.

India reminds me of Lagos – dirty and poor, but with more people, older cars and lots of three-wheel scooters. Madras was not too bad, but the bit I saw of Bombay was terrible, worse than Lagos.

India has the most overstaffed businesses in the world; wages are low and there are a lot of them. We went to a fabric store. It was a building with five storeys (each about 200 square metres) and about a hundred employees. It had thousands of different colours of fabric. The ladies in India have it much easier than their Western sisters when it comes to shopping for clothes, as they all wear saris and there is no decision about style; for size, they multiply their height by their waist to get the length of fabric, so all that is required is to choose the colour.

India is very security-aware; there are soldiers and armed guards everywhere. They search your baggage four times before you board a plane.

Needless to say, I got ill in India, and the problem in India is that you have to puke into toilets that you don't even want to stick your arse into. The more I travel, the more I see that there are some peoples who are just not happy until they have a dirty toilet.

Annabelle made the observation that Lagos and Singapore are the most opposite of countries, so using this as a scale, with Lagos at zero and Singapore at ten, I would slot India in at two or three, Indonesia at five and SA at seven or eight.

The judgement criteria would be organisation, standard of living, corruption, infrastructure, education and discipline of society. So unless you want to stay in an over-controlled, brainwashed society, SA comes out best. There are, of course, other countries that would also rate well.

Poor Inez had only been in Lagos for four weeks when I was off for three. She braved her way through it, and thanks to friends; you helped.

We are settling down and getting some routines. We work very long hours (fifty-six-hour week) and free time is limited. Cost of living is high and it is difficult to get anything done quickly.

We will be in SA over Christmas.

Please write; it is nice to hear from friends and stay in touch.

Cheers and miss you all,

Arthur

Letter from Nigeria Seven

Transport

September 2004

Well, it is now six (6) months that I have been in Nigeria. One more week to go, and it is six months that I have not smoked.

As promised, this letter will deal with transport in Nigeria. It was more difficult than I expected to get photographs, as generally the local population is uncooperative and they want money to have their picture taken. (Recently some South Africans were put into jail for a couple of hours when they photographed a taxi and the driver, with his police mates, got difficult.)

A practice here (which I have also seen in India) is that you pay to take a camera into certain places, and the fee increases as the camera becomes more sophisticated. A digital camera can cost up to R150. I presume, as there is little else they have to offer or sell, that this is one way of generating an income. I don't understand the logic, as it presumes that if I take a photograph, I won't come again. Imagine if they had to impose this at the Kruger Park, or if we had the same attitude to sex; take a photograph of her and the spouse is redundant – I forgot, somebody has to wash the dishes.

Transport in Nigeria can be best described as a mess, and if you suffer from road rage, it is best to stay away. There is a railway network that serves a limited part of Lagos; however, only a fraction of it is operational, and then only sporadically (delays of up to four hours are the norm).

Road transport is slow, frustrating and hazardous due to the following:

- Poor road maintenance.

- Unnecessary roadblocks that the police use to extort money.

- Badly maintained vehicles where only the hooter is guaranteed to work.

- Totally undisciplined society.

The hooter is called a 'trafficator' and it's not used to warn of hazards but rather as a communication medium and sometimes as a chill pill to relieve frustrations. Indicators are optional fittings on a car and are never used; if you are passing somebody, you hoot at them to advise. You can spot a new expat South African because each time somebody hoots at his vehicle, the blood boils, accompanied by the usual hand gestures and cries of 'Fuck off.' I have explained very nicely to my driver what the hooter is for and when he may use it, and there are no exceptions.

The ill-discipline is a disgrace and really amounts to pure selfishness. If a taxi wants to offload passengers, it will just stop in the middle of the road and make no effort to pull over. When there is a traffic jam or when the traffic is moving slowly, people will drive on the pavement or even on the wrong side of the road, endangering all road users.

There are a multitude of informal stores and stalls lining the roads and they encroach onto the road and invariably slow down the traffic. I presume there must be a marketing advantage by having your stall in the middle of the road. I have seen this in the Transkei, a rural area of South Africa, as well, and in weaker moments have had fantasies of being a tank driver.

The next photograph shows how merchants use the road as part of their stores; it also shows the type of shops and goods available (note the new TVs).

In the photo below, the guy was not too happy about me taking his picture and wanted about US $3.50 from me. I told him to get lost. Who does he think he is, Michael Jackson?

We each have drivers allocated to us who are really just chauffeurs, as when they are not driving us or washing the cars, they are supposed to do nothing, which they do remarkably well, except when they are sleeping. When drivers are employed, as part of the interview, they have to prove that they can sleep in the most uncomfortable places prior to being appointed.

No vehicles are manufactured in Nigeria, and Europe uses Nigeria to dump all its unwanted vehicles (in many European countries, cars older than three years need to pass a roadworthy test each year). To make matters worse, Nigerians don't believe in maintenance, so generally the vehicle will be driven in ever-worsening condition until it stops, and there it is left to become a fixture.

The most common type of public transport is the minibus taxi, and if ever you thought the ones in South Africa were obnoxious, dangerous and absolute hooligans, these are worse.

Most are manufactured as twenty-seaters but used to carry over thirty. They all have conductors, and if you look carefully at the picture above, you can see, through the windscreen, the conductor hanging out of the door. I believe these conductors only get paid a commission and will squeeze as many people as possible into the taxi and take their place hanging on to the outside. A colleague once referred to them as *'die naaste ding aan 'n bobejaan'* (the closest thing to a baboon). He was having a Lagos Blues day.

Large buses are used for travel between cities and are generally considered by the locals to be relatively expensive. I have seen passengers sitting on the roof. I believe there is a bus service from Lagos to South Africa that takes seven days non-stop, and passengers are expected to bring their own food. I must say, if I had to be stuck on a crowded bus in the tropics for seven days and seven nights, I am sure I would also be sitting on the roof.

Sedan taxis are also available but considerably more expensive than the minibus. Note in the photograph below the Switzerland identification sticker – no prizes for guessing the origin of this car.

The cheapest and most convenient mode of transport is the Okada motorbike taxi; however, it is also the most dangerous. Okada drivers are notorious hooligans.

The picture above is of a typical Okada motorbike – note that he has fitted extra-large hooters, and the one at the back has bent his handlebars up, so as to weave between smaller spaces in the traffic. What the photograph does not show is that both front brakes have been disconnected. They will carry anything, and I have seen passengers carrying two TVs, a 6 metre drainpipe (which nearly made kebabs out of Inez and me), towing a lawnmower, and carrying a bale of straw. In the following picture, you will see, if you look carefully, that there are four Nigerians on that Okada.

Most Nigerians just can't understand why I burst out laughing every time I see a 'Mary Poppins' as above.

Most Okadas come from China and are 125 cc and, when new, cost about US $650.

There is a very easy and quick method of dealing with car accidents. Instead of expensive insurance, reporting to the police, filling out lengthy and complicated forms and expensive court cases, here in Nigeria the two drivers square up with their fists. Needless to say, the innocent party is often the bigger one of the two, and normally the claim is settled there and then either by payment or the annexure of the vehicle.

A colleague of mine who has spent some time in Angola says that there the first policeman on the scene of the accident would hear both parties' accounts of the incident, then determine who the guilty party was, and how much should be paid to the innocent party and to himself. No doubt perceived wealth helped determine the guilty party.

Travelling in Nigeria is difficult, and I have noticed that anger, frustration and most of all depression are often linked to the amount of time people spend travelling by car, as during every car journey, one experiences the following:

- Appalling animal cruelty.

- Terrible human suffering.

- A complete lack of respect for fellow man.

- Endless window-shopping through a car window.

When I arrived in Nigeria, a South African expat who had been here for a while described living in Nigeria as stressful, and I now understand why.

The sad part is that things are not going to improve, as Nigerians tend to live in a fool's paradise and always say that things are great in Nigeria. It saves them having to do the things necessary to sort it out. Below is part of an article taken from a local newspaper paying tribute to the repair of a road, which was a relatively simple project, and which they did not do well, as the levels are wrong and when it rains, it still floods. Most of the roads in the same area (main business district) are in a poor shape and 4x4s are needed when it rains. As my scanner is also a victim of Bill Gates's upgrades, the article has been retyped verbatim.

Motorists who ventured into this once terrible but important road, a year ago had horrible tales to tell. The nightmarish experience extended to pedestrians, workers and business owners whose offices are located on Adelola Odeku Street, Victoria Island. A driver then may well have been required to visit a mechanic. The situation was made worse after any downpour. When the situation could get no worse, Lagos State government contracted construction giant, Julius Berger to repair the deplorable road. As, Julius Berger took its time to do the work, business groaned as potential customers shunned the area. Several businesses admitted to lost sales amounting to about 20% of turnover, but things have now changed. Mr A Niran, the floor manager of Park 'n' Shop said prior to opening of the new road sales were dull. 'Then a lot of our customers had a hectic time coming to shop because of the deplorable state of the road. Most of them particularly those who came in their cars often complained of the damage to their cars due to the poor state of the road' Niran said. 'The opening of the road has encouraged shoppers to come in droves. Frankly speaking, the rehabilitation of the road has largely increased sales. More and more people now come to buy their daily needs', he enthused.

In her view, Angela Nkiru, an attendant at Rocky Telecoms located at Rocky Plaza, also said that the reconstruction in the road has not only increased patronage but also encouraged prospective customers to visit the plaza. 'It was quite a terrible experience to us because we almost lost our customers who complained of the poor state of the road. In fact, most of them stopped coming to the plaza rather they preferred placing their order through the telephone'.

As a result of this, Nkiru said in order to generate sales, staff had to market their products by visiting the customers in their offices and in their homes adding that the poor state of the road accounted for the drop in sales. Asked if they would have relocated from this street, she said: 'Not at all because it would be difficult to get a place like Adelola Odeku because it is the hub of business and commercial activities in Victoria Island. For example there are several banks on the road as well as telecommunication like Globacom, Multi-links, Linkserve and Coscahris Motors which sells BMW and Range Rover brands'.

Nkiru said that the perennial traffic on the road had also eased considerably thus creating adequate access for motorists and pedestrians who ordinarily would avoid the road due to its deplorable condition. 'Apart from this there is adequate parking space for customers due to the dualization of the road', she elaborated. Some roadside traders who depend on heavy traffic to offload their wares including GSM corner stands and GSM recharge card vendors told Island

> *News that since completion of the road, commercial activities has increased as it is very much convenient for motorists and pedestrians to pass through the road.*

Besides the poor writing, the journalism is a disgrace, as the reasons for why it took so long and the possible plans for the rest of the neighbourhood are not addressed at all.

Lagos is a coastal city at the mouth of a large lagoon, which forms a wonderful natural harbour, with access by water to many parts of the city. This facility is horribly under-utilised, with very little water traffic.

The plastics plant (the project I am here for) is situated on the coast and will have its own harbour; thus, a geological survey ship was hired to take soil samples. We were invited to visit the ship to look at the work they were doing and to have lunch. The ship was anchored about 2 km off the coast and about 70 km east of Lagos Harbour; hiring a launch from the harbour was expensive, so the local fishermen were hired to take us from the beach to the ship. The pictures below show the locally manufactured Lagos speedboat and how it is sealed. It leaked like hell and the driver spent most of the time baling water. In the background is the ship. A rope ladder was hung out for us. The sea here is extremely rough and the seabed is extremely steep (swimming is impossible). This ride was scarier than any rollercoaster I have been on.

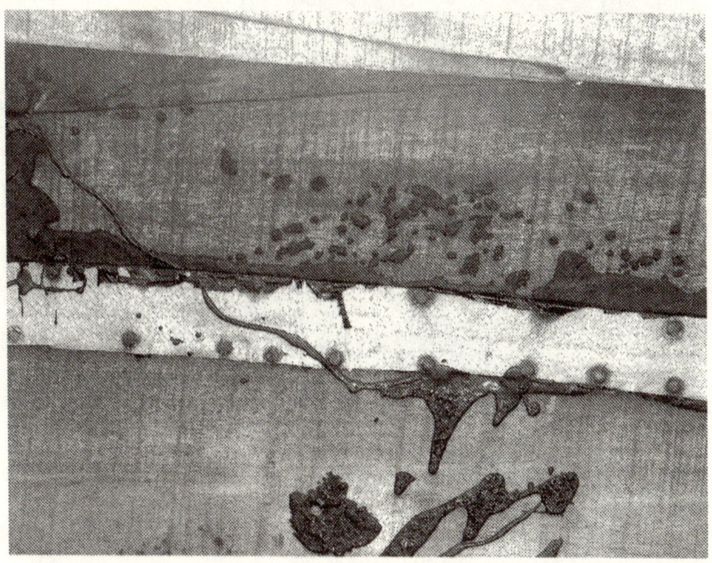

The Lagos area (climate, terrain and waterways) reminds me of Florida and has a massive potential for marinas, golf courses and holiday resorts.

About 50 km west of Lagos is the port town of Badagry, which was where the slave trade took place. It was here that the slaves were purchased from the local chiefs and loaded for shipping to the Americas. The town is listed as one of historical importance, so being good visitors, we went off for a visit.

Five hours there and back (it was on a Sunday, when the traffic is supposed to be light), and after a visit to greet the local chief, we needed less than an hour to see all there was to see.

I have twice been to the world's largest museum, the Smithsonian in Washington, and now I have also been to the world's smallest museum, the Mobee Family Slave Relics Museum. It consists of one room about 15 square metres with about half-a-dozen artefacts and a few articles on slavery copied out of the school library encyclopaedia.

In the photo above are Inez and a friend of mine, Janus, standing in front of the museum. The building is of no historical significance and is just a local shack.

Inez and I are settling down as much as we can and getting into a routine. We are still eating loads of fruit salad, walking miles and swimming lengths and lengths. I am now playing squash once or twice a week and Inez has started tennis lessons (they are very cheap here and the quality not bad).

In three weeks' time Barbara (my sister) and her family will be visiting us for a week. Inez and I are both looking forward to the visit, as we miss our family and friends and especially all the nephews and nieces. Both of us will be working, but will get some time off to take them around, and we will have the evenings and the weekend to catch up with South Africa and the family gossip.

In November, Inez and I are going to the UK to watch the rugby. We will be joining Peter (my brother) and his wife and will watch the tests at Twickenham and Murrayfield and spend the week between travelling around by car and will be seeing some old friends. I am bracing myself for a big party.

We will be in South Africa for Christmas and New Year. So the rest of the year is going to be busy, but something to look forward to.

Thanks to the subscribers who wrote and to those who gave assistance with solving my photo-editing problems.

Cheers to you all,

Arthur

Letter from Nigeria Eight

Shopping

October 2004

This letter is about a week late; however, as many people are more than a month late with their responses, I don't feel so bad. Please remember, subscription fees are one return letter from every reader. To those of you who did respond, thanks very much, and to those who are now feeling guilty, the best cure is to quickly write two letters. Inez and I do enjoy hearing from you guys all around the world.

As promised, this letter has to do with shopping in Nigeria, but I also have a section on how electricity is managed in Nigeria. I apologise to those who are not that technically minded, but I have tried to look at the human-interest side.

Cast your mind back to the good old days when you would go shopping, taking a whole lot of notes in your wallet to the shops, and as your wallet became lighter, you knew that shopping time was over. Today you all use little plastic cards that allow you to buy anything you want, so there is no end to the shopping; you just worry about it later. In Nigeria, we do have credit cards; however, use it once, and within half an hour, half of Nigeria is also using it (and as there are 140 million of them, it gets quite expensive). The same applies to autobank cards. Thus, in Nigeria, we are still on the hard currency system. The problem here is that things are expensive and the note denomination is not all that big, (I believe this is to prevent counterfeiting).

In the photo below is our allowance for one month, which is just over US $1,000 (and Inez's pay packet). So as we pay the maid, DSTV (SA satellite TV), etc., the pile gets smaller, and the trick here is to make sure that you see the end of the month with something still in the pile, because you can't just go to the auto bank and draw on your overdraft or use the credit card. The same applies to going out – when the money is finished, you go home, no 'one for the road'.

This reminds me of my sister's late father-in-law, Dick Taylor, who was an engineer to the core and believed that one did not need accountants, as all you need is a bag with all the money, and when the level is low, you slow down on spending and pick up on sales and vice versa.

An analogy in the United States would be like paying for your groceries with US $1 bills. As you would expect, our money-counting skills have improved.

Generally the money is old and smells, and the other day when I saw a woman hiding money in her bra, I began to understand why. (I don't even think about the other places it could be hidden.)

We have supermarkets in Lagos, and the picture below shows the shop Park and Shop, which is also commonly known as 'Park and Rob'. It is very much like a big Spar;[7] it has a lot of imported goods, mostly from Europe, the Middle East and South Africa. It is Lebanese-owned, and one can clearly see that influence. We do find items not normally available to us in South Africa (e.g., 500 different makes of olive oil). The cost of goods here is about two to three times that in South Africa.

Shopping here is much like shopping at Pick 'n Pay (South African Tesco or Wal-Mart) in South Africa. It only takes twenty minutes, and the wife and I argue and we have a trolley full of things that I don't want, (e.g., humus) and things she doesn't want (e.g., tinned Viennas [hot dog sausages]) and things neither of us want (e.g., huge box of soap powder because it's on special).

---

[7] Chain of small supermarkets in South Africa.

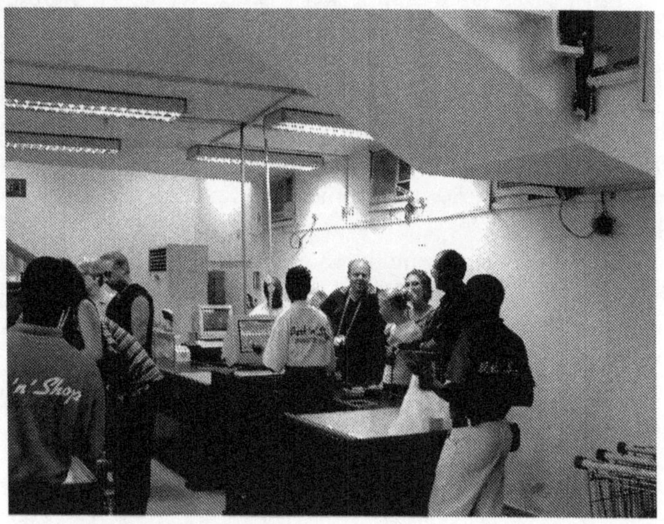

Taylor family at the till coming to grips with the money

Lekki Market

The other place we shop is Lekki Market. We generally restrict ourselves to dry rations, fresh fruit and vegetables, general household goods and local arts and crafts. All the stalls are owned by individuals who have two objectives – first to get you into their shop and then to excessively overcharge you. No goods are marked, and when you, with a white face, enquire about a price, you are quoted a price three times too high and the negotiation process begins.

The problem is, during the above process, all the surrounding shop owners desperately market their goods to you, which involves either grabbing you by the arm and pulling you over to their store or sticking their merchandise in your face.

Buying something simple like a kilo of potatoes entails the following:

- finding suitable potatoes,

- negotiating the price down to one third of the initial asking price,

- then stopping him from taking out his rotten potatoes from the back,

- threatening his neighbour that if he sticks that pineapple in my face again, I will stick it somewhere else, and guess which side first,

- not minding that the potatoes were not weighed and you have taken his word for it that it was actually one kilo.

Fish is also sold at Lekki Market. I believe the way to eat the fish in the photo below is to get yourself within dashing distance of a latrine, scoff it down, and take off.

A lot of arts and crafts are sold at the market, including ivory. I refuse to buy any, but I have noticed the Indian and Chinese expats buy it in large quantities. So much for the worldwide ten-year ban on ivory trading.

A number of stalls have paintings, and fancying myself a local patron of the arts, I decided to buy some. At the first stall I went to, I was shown about three hundred paintings, varying in price between US $15 and US $50, and the salesman was claiming to be the artist.

I asked him how long it takes to paint a picture. Two weeks to a month, he replied which means he has spent most of his life painting. I also noticed that most of the paintings had different artists' names on them, so for all I knew he could have raided the local primary school's art classroom for stock. The painting I bought also had a dramatic price reduction.

In the photo below is the area we call Mica, as it is where we get our hardware.

The picture below shows the local Tiger Wheel and Tyre (South African tyre company).

The range of tyres also includes badly worn second-hand ones.

Below is a butchery selling bush meat. These animals here are called grass cutters and are considered a pest. They can destroy a crop in no time. To me they look like overgrown rats, but apparently they are a type of rabbit. Either way, I will not order it in any restaurant.

The picture below shows OK Bazaars[8]-type marketing at its best, although I am not sure if it is appropriate for property. The grammar is worse than mine on a bad day. I believe that when buying property in Nigeria, the biggest problem is making sure you buy it from the owner.

---

[8] OK Bazaars is a South African chain of stores, similar to Woolworths in the UK.

In Nigeria, power failures are daily occurrences, and we all have generators and UPSs. The reason for it is two-fold – namely, a shortage of working generating plants and a very poor distribution infrastructure. The moment it rains, the lights go out.

In the picture below is a typical example of what an electrical board looks like.

This installation above is at a friend's house, and one evening when he did not have power, I went to assist. It was quite a daunting task, as it had been raining, and I was standing in about 2 cm of water, and needless to say, all the fuses were bridged out with wire, and not a single earth was connected.

Lately in the course of my work, I have been inspecting and approving electrical installations, and have noticed that the electricians tend to coil all the conductors (as in the above picture). I enquired from an electrician about why they do this, and he explained the reason is that if the conductor burns off, there is plenty of slack to reconnect it. I rest my case.

In the next picture is something we see all the time here. The transformer is live, and at the top of the red cables, there is no insulation. The voltage is 11,000 volts – touch it and you are fried.

A colleague of mine was at a bar with an outside veranda one night, a place well known by its reputation. A storm was brewing and the winds were picking up. As the rain started, tree branches caused the overhead power lines to touch and short out. Due to inadequate protection equipment, the fault sustained for about thirty seconds (in layman's terms, it sparked for a long time). During the fault, the overhead lines melted, resulting in a glowing waterfall of molten steel flowing to the ground. When the lines had finally melted away, the sparks stopped and all the lights went off. The reaction from both the locals and expats was to stand up and in chorus sing 'Happy Birthday'. By the end of the singing, the diesel generator had started, the lights were back on and the party continued unabated. Now we are not quite sure if this was a spur of the moment reaction or a normal way in which the people of Lagos cope. Surely this could only happen in Nigeria.

My sister and her family came to visit us. It was great to see them and catch up on news from home. To all those who sent presents and essential supplies, thanks very much; it is really appreciated.

We did manage to take them around Lagos, and although Lagos has no particular tourist attractions and travelling in the city is slow, exhausting and sometimes disturbing, the Lagos experience is unique and impossible to describe in words.

The picture above is the gang at a barbeque outside our apartment. It looks like we are in the bathroom, but that is just because they put tiles on the outside walls of the building.

They used the holiday to relax by the pool, catch up on some movies and spent the evenings partying with us.

Next week we are off to England and Scotland to watch the rugby with Peter, my brother, and his wife, Marilette, and to meet old friends.

The origin of this tour is that Peter and a friend bet against the Springboks and lost. The penalty was that they had to see South Africa play against England at Twickenham, and we are joining them. We are looking forward to a week of seeing friends and family, rugby and drinking copious amounts of the local brews and experiencing civilisation again.

We are still scheduled to be in South Africa over the Christmas/New Year period.

Remember, subscriptions are due, and there should be another letter before Christmas.

UP THE BOKKE. (Up the Springboks.)

Cheers, love and miss you all,

Arthur

Letter from Nigeria Nine

Travelling in Europe

November 2004

I am finally back in Nigeria after twenty days in the UK and France.

As you all know, my brother and a mutual friend had an agreement that if South Africa won the Rugby Tri-Nations, they would go and watch them play at Twickenham in London in November. I am not sure how it happened that Inez and I joined them and, if we were coerced into anything, it was without any resistance and with no regrets.

Although we had confirmed bookings on Virgin Atlantic direct from Lagos to London, I, by chance, found out at a later date that we longer had them. After few nervous days during which I threatened to throw the travel agent and his staff out of his office window, we finally got tickets on Alitalia via Accra and Milan to London.

When flying, there are different classes, which start at First Class, then Business, Economy, Baggage, Strapped to the Wing, Alitalia, and finally Alitalia to Nigeria Class. The aircraft was filthy, especially the toilets, and almost no food was served.

Although we left Lagos early, we were delayed in Milan, and we had to go direct from Heathrow to the match and only arrived five minutes before the game started.

In spite of:

- South Africa getting a rugby lesson from England;

- it being freezing cold;

- Inez having a tray of beer poured over her by accident;

- spending half an hour after the game looking for our wives, only to find them sitting in a pub.

The Twickenham match and after-party were a great experience that will remain with us for a long time.

After the rugby, Inez and I travelled up to Scotland with my brother, Peter and his wife, Marilette. Along the way we stopped to visit friends, and did the tourist thing in York. It was great to see old friends again and to spend time with Peter and Marilette.

From the outset, Peter had a thing about the English town names, and by either changing a letter or the pronunciation, he could come up with something quite humorous. I did my best to convince him that a town name does not have to end with a 'dorp' or 'berg' and was slowly making progress when we came across the village 'Knucksford'. I gave up and admitted defeat.

In Scotland we went to see South Africa play Scotland at Murrayfield, where they performed a bit better (we will neglect to mention that currently Scotland is not considered to be a strong team).

Marilette, Arthur, Inez and Peter

Many of the Scots wore kilts to the game, and besides the lily-white legs (it looked as though they had been dipped in bleach), these must be some of the toughest men in the world, as it was ice cold, and no underwear or stockings are worn with a kilt.

After the Murrayfield game, Inez returned to Nigeria, and I met up with two colleagues for business meetings with suppliers. The meetings in England were fine and finished early, so I managed to spend a day in London.

Then it was off to Paris. It has been some time since I was last there, and it looked a lot cleaner and more organised than I remembered, but Paris is still Paris. An example of this is when we got to Charles de Gaulle Airport, many French men and women were smoking, in spite of numerous signs showing it was a no-smoking area. My colleague decided to have a smoke, and as soon as he lit up, everyone in authority came dashing over, telling him to stop.

On the Saturday afternoon, we were invited to the 'Stade de France' for a motor-racing event inside the stadium, where they have built a track. A number of drivers from different countries and of different motor-racing types were present, and the organisers had them competing in a variety of cars, from the world-champion rally car to a Ferrari 308. Amongst those present were Michael Schumacher, David Coulthard, Felipe Massa, Jean Alesi, the American cart champions and quite a few rally drivers. Generally the rally drivers were faster than the Formula 1 guys. As the stadium was very small and compact, we were able to see the drivers up close. And to Carol, Barbara and Inez, I can now confirm that my butt is cuter than David Coulthard's. It was a great evening, and Stade de France is a very nice stadium; however, it was ice cold, and it took me hours to warm up again.

On the Sunday, my colleagues left and I remained behind in Paris for two days, awaiting my flight back to Nigeria. I spent most of that time walking around Paris and celebrated my birthday by drinking a couple of beers while sitting alone in a pavement café.

The return journey was via London and Milan, which was tedious at best and took about twenty-eight hours.

When leaving from Charles de Gaulle Airport, I went to the

departure lounge with the intention of getting a meal. To my shock and horror, I discovered that the best I could get was a stale sandwich and a beer. The Frenchman behind the counter was not impressed when I told him that it reminded me of Nigeria (remember my incident at Lagos Airport).

As described earlier, Alitalia is a mess, and they lost some of Inez's baggage the previous week. So when I boarded my flight in London, I tried to get all my baggage as hand luggage, and I used the excuse that as they had already lost one of my bags, I was not having them lose another. After a long argument, I was given two options: either I fly with my baggage in the hold, or I stay in London. So I relented, but was peeved, as I was sure when I got onto the plane to Nigeria, the Nigerians would have so much hand luggage that the aircraft would look like a South African taxi at the beginning of Easter weekend.

To my surprise, Alitalia was serious about it, and determined that every person travelling would only have one piece of hand luggage. The Nigerians are streetwise, and when they check in, they don't bring all their hand luggage with them, so when they climb on the plane, each person has about four suitcases as hand luggage. Not to be outdone, the Italians decided that as each person boarded the bus to the plane, they would remove the excess hand luggage. The Nigerians responded by refusing to hand over the same, and then the circus started. Picture this: the Italians trying to organise a bunch of Nigerians. Every person within Alitalia associated with baggage decided to throw their weight around and order people to give up their luggage. This included the baggage handlers, the bus drivers and the ladies checking the tickets (the noticeable exception were the police and security – although they were smartly dressed, they stood to one side and watched). The first time they tried to remove the luggage from a Nigerian, he promptly ignored them and climbed onto the bus. The next six passengers did the same. The Italians responded by closing all the doors and summoning the police. Half an hour later, two very smartly dressed policemen with white holsters and belts and caps appeared, and eventually the unnecessary bags were removed. We then proceeded to board the bus one at a time, with numerous arguments and fights. The best I heard was a lady claiming that her baby's food was in the bag, and without that food, the baby would die. The kid was about two years old and was running around making a nuisance of itself. Needless to say, we were an hour late taking off; again there was very little food, and even the water was rationed by the glass.

I have noticed that every time I have boarded a flight to Nigeria, the aircraft is the oldest and most grotty one in the fleet and is invariably parked in the far, distant corner of the airport. But reviewing the behaviour of both Alitalia and the Nigerians, I have come to the conclusion that when it comes to Africa, the West doesn't care, and the Africans don't really mind either. Being an African, this is depressing.

To all our friends and family, a merry Christmas and a prosperous New Year. Many of you we will see over Christmas. Letters will continue in the New Year, depending on the number of replies we get.

Cheers and love to you all,

Arthur and Inez

Letter from Nigeria Ten

Coming Home

January 2005

Hi to all subscribers (paid up and those in arrears),

Since the last letter, we have been to South Africa and have seen many of you. We had a wonderful time, ate enough decadent foods to last us another year and drank a huge volume of beer. Last week I had my blood pressure checked again, and it's the lowest it has been for more than a year, and I attribute this to all the beer I drank whilst in South Africa (I shared this opinion with my doctor, but he just shook his head).

The success of the holiday can largely be attributed to many people taking time off to see us and particularly to people who helped with organising, hosting and opening their homes to us, namely Alice and Bob, the Taylor family, the Carlisle family, Lida, the Lucases, Tom and Di, Maureen and the Deken family.

When returning to Nigeria, our Nigerian adventure started at Johannesburg Airport, as when checking in, we were informed that our flight had been delayed for four (4) hours due to heavy dust at Lagos Airport. The dust cloud at Lagos, called the Hamatan, is an annual event that lasts for a few weeks and originates from the Sahara Desert. The dust can get so thick that no shadows form and visibility is limited to less than 100 m. The dust is not red, as one would think, but is rather a bland gray and makes the world look dull. The photos below were taken directly into the sun at 09.00 in the morning and gives an indication of the thickness of the dust.

Dust in the morning sun

The delay of the flight was a result of Lagos International Airport not having radar or other navigation equipment; they do have it, but it is not operational. We were lucky, as our delay was only five hours. The Friday flight left on time and flew to Lagos but at the last minute was diverted to Accra in Ghana (about 400 km west of Lagos), refuelled and returned to South Africa. The next day, again a flight took off from Johannesburg but halfway up the coast of Africa turned around, refuelled at Luanda and returned to SA. Eventually the flight landed in Lagos late on Monday night.

A colleague of mine was on the Wednesday flight and the pilot twice had to abort the landing at the last minute before eventually completing it.

I believe the Hamatan is relatively bad this year; however, it should come to an end soon, when the rain begins. Due to the dust blocking the sun, it is a lot cooler, and we have had temperatures in the low 70s F, making it quite pleasant. The temperature was never cold enough to warrant wearing a jersey, but many wealthy Nigerians who had spent time overseas and had acquired expensive coats used the opportunity to show them off. Those of us who suffer from hayfever do struggle.

So, relatively speaking, we were not too badly off, in spite of arriving in Lagos at 1.00 in the morning and eventually getting to bed at 3.00.

The day before we left for South Africa, a water pipe in the kitchen wall burst, and Inez convinced our landlord to replace the cupboards while carrying out the repairs; this was all to be done while we were on holiday. Needless to say, the work was less than half complete, and we returned to a construction site in our apartment. So with the dust from the Hamatan and the kitchen, we had our own little Sahara in our apartment.

Below is a picture of our 'new' kitchen. They have strange ways of building here. The cupboards are imported from Italy as complete units, and the tabletops are cast concrete slabs supported on masonry walls. I am sure I could have done a much better job at half the cost.

The wine rack was an idea of Inez's and a friend and not bad for US $7 material. We had a bare wall, and Inez came up with the idea of having a graffiti wall. All our friends are too shy to write their names on the wall, so we are organising a party, and hopefully after a few drinks they will lose their puritan upbringing of not writing on walls and their sense of humour will be stimulated. I will show you the wall when it's suitably vandalised.

In this letter I will not deal with a particular theme, but rather just give some isolated items on Nigeria.

In the photo below is something you see fairly regularly in Nigeria. The registration of title deeds is lacking, and when buying property, it is very difficult to establish if you are in fact buying the property from the owner and not an impostor. Nigerians are well known for their scams. The modus operandi is to identify an attractive property (needless to say, it is not their property), set up scam offices that include lawyers, owners and estate agents and then to sell the said attractive property to a number of unsuspecting buyers. The reference to 419 originates from the clause in the Nigerian Penal Code, which prohibits deriving an income from fraudulent transactions.

I have heard it said that of all the cars manufactured Land Rover has the most still on the road. If you ever wondered what happened to them, they end up in Nigeria carrying building material. There are thousands here in various states of repair, and I think their popularity is due to these vehicles being relatively simple to maintain and strong enough to carry the loads on the Nigerian roads.

The next photo is a Land Rover graveyard – note the different applications; e.g., the breakdown.

The photo below was taken on a plot of land just behind our hotel. I thought quicksand was a Hollywood myth. I was tempted to go have a look but would feel awfully stupid if I were caught in it after I was warned. A further danger, which is real and probably worse, is that the local builders use the plot as a toilet.

While on this topic, there are many toilets in Nigeria that still use the bucket system.

I am not sure if the owner of the truck in the photo below has a sense of humour or takes his work seriously.

This photograph shows the chaos of Lagos. It is taken in the CBD after a thunder shower. Note the dress code. Suit jacket remains on but shoes are removed and pants are rolled up.

When we were in South Africa, we made our pilgrimage to Exclusive Books[9] to stock up on reading matter. The bookstores in Nigeria are poor and the books are limited to sentimental religious bumph or ten years out-of-date MBA textbooks. Whilst in Exclusive Books, I saw a news magazine (the name escapes me), that focuses on Africa. The edition on the shelves was focused on marketing Nigeria for a post on the UN Security Council. I personally believe that the Security Council does need a review, and can no longer have its members limited to the larger economies of the world and needs broader representation. However, how can Nigeria want to solve world problems when they cannot even get the basics right in their own country? I am referring to issues like the high infant mortality rate, a police force that does nothing but extort money from the public and industry,

---

[9] Chain of bookstores similar to WH Smith or Barnes and Noble.

being the fifth-largest oil producer and still having to import all its refined petroleum products. I hope that my experiences of the last year have not made me more cynical, but I am sure that the drive for Nigeria to become a member of the Security Council might well be for certain individuals' personal gain.

Some of you might know that I was due to have a business trip to Estonia in January; however, this has been postponed till probably early March. I believe it will be warmer, but in my book, the difference between 14° and 23° Fahrenheit is very fucking cold and just fucking cold.

I have been asked by a number of people if they may resend my letter to friends and family, and I have no problem with this. However, it does not get any of you off from paying your subscriptions.

Thank you to all those who assisted with photos and other contributions.

Below is what memories are made of, on holiday with the family over Christmas. The star is my niece Caylin and getting the love is yours truly.

Cheers, love and miss you all,

Arthur

Letter from Nigeria Eleven

Wedding

February 2005

This letter deals with two topics: weddings and corruption. Please be assured that the topic selection is arbitrary, and those of you who are struggling to distinguish between the two, let it be on your conscience.

Many of you may remember Inez's and my wedding and the following series of incidents: the Rolls-Royce broke down between the church and reception; the silver knife broke while cutting the cake for the photo; an untimely thunderstorm (breaking a seven-month drought) came and soaked half the guests; the band refused to play, as some of their equipment was wet; we were given the keys for a honeymoon suite that was already occupied; we had to enter the real honeymoon suite through the window.

In spite of all this, we still had a great party; hence, I now have the view that the main objective of a wedding should be a big party and should take place in two phases – first an extended lunch break on Thursday or Friday to the Registry Office, and secondly, a big party on the Saturday. No need for dressing the bride up in a funny ball of white lace, having hours of long, boring speeches and the inevitable waiting for photographs to be completed. A colleague of mine decided to get married, and did not follow my advice, but still invited us to his wedding. The wedding was held in the suburbs of Lagos.

Suburban Lagos

The service started at 11.00 at which time there were about twenty of us in the congregation – in true African style, most people arrived late. At the end of the service (three hours later) there were more than two hundred people present. Initially things were stiff and slow. The sound system and singing were terrible, and the hymn 'O Lord My God' was painful to listen to. Initially the service more or less followed the traditional 'Anglican' order of mass. Once the marriage vows had been said, the pace and style of the service changed and things began to rock. The singing and music improved, and the congregation started to dance. The next picture shows the bride and groom jiving down the aisle on their way to sign the register.

We, the congregation, stayed behind and continued to sing and dance until their return.

The next two hours of the service were far more evangelical in nature, with some of the performances outdoing Billy Graham at his best.

There were seventeen priests (note the number of women) who officiated over the service, with a further twelve in the choir, four in the band and ten ushers. All the pastors played a role as translators. The service was conducted in English as well as a local language, Yorubu.

Below is an extract from the programme – note the size of the bridal train and the spelling mistakes. This is how they talk, and no one owns a dictionary.

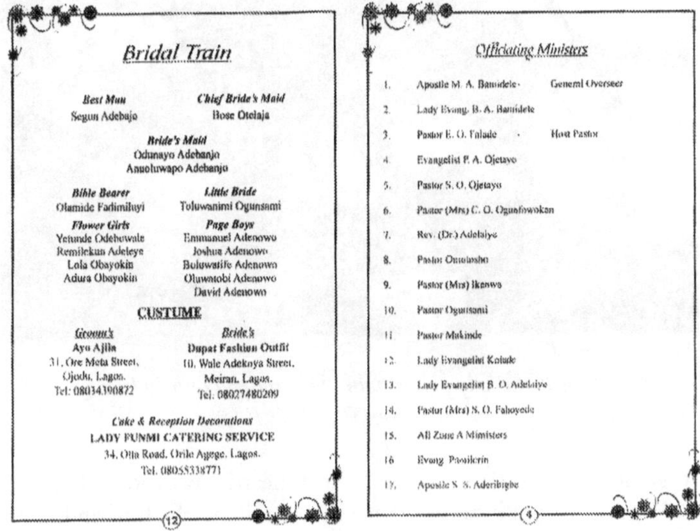

Bridal Train

Best Man	**Chief Bride's Maid**
Segun Adebajo	Bose Otelaja

Bride's Maid
Odunayo Adebanjo
Anuoluwapo Adebanjo

Bible Bearer	**Little Bride**
Olamide Fadimiluyi	Toluwanimi Ogunsami

Flower Girls	**Page Boys**
Yetunde Odehuwale	Emmanuel Adenowo
Remilekun Adeleye	Joshua Adenowo
Lola Obayokin	Buluwatife Adenowo
Adura Obayokin	Oluwatobi Adenowo
	David Adenowo

CUSTUME

Groom's	**Bride's**
Ayo Ajila	Dupat Fashion Outfit
31, Ore Meta Street,	10, Wale Adekoya Street,
Ojodu, Lagos.	Meiran, Lagos.
Tel: 08034390872	Tel: 08027480209

Cake & Reception Decorations
LADY FUNMI CATERING SERVICE
34, Otta Road, Orile Agege, Lagos.
Tel: 08055338771

(12)

Officiating Ministers

1.	Apostle M. A. Bamidele	General Overseer
2.	Lady Evang. B. A. Bamidele	
3.	Pastor E. O. Falade	Host Pastor
4.	Evangelist P. A. Ojetayo	
5.	Pastor S. O. Ojetayo	
6.	Pastor (Mrs) C. O. Ogunfowokan	
7.	Rev. (Dr.) Adelaiye	
8.	Pastor Omolosho	
9.	Pastor (Mrs) Ikenwo	
10.	Pastor Ogunsanti	
11.	Pastor Makinde	
12.	Lady Evangelist Kolade	
13.	Lady Evangelist B. O. Adelaiye	
14.	Pastor (Mrs) S. O. Fahoyede	
15.	All Zone A Ministers	
16.	Evang. Pasailerin	
17.	Apostle S. S. Aderibigbe	

(3)

Wedding programme

The pastor who delivered the sermon looked, sounded and behaved like Eddie Murphy, and I struggled to keep a straight face. It felt as though I were getting a sales pitch from a con man. The core theme of the sermon was that the wife must be subservient to her husband, culminating in the statement that the man should not oppress his wife. He must just see, listen, think and speak for her. Evidently the only thing that he and I can agree to is that we have very different definitions of the word 'oppression'. At the end of the sermon he gave some 'fatherly' advice to the bridal couple, which was in the form of a list of 'things to do': sit together, walk together, sleep in the same bed, eat at the same table, and when he got to bathe together, I nearly jumped up and shouted, 'Hallelujah every day, hallelujah.' It was 90° Fahrenheit, 90% humidity and we had been in the church for two hours.

One of the lady pastors with the same bad taste in hats
as Queen Elizabeth II

It is not the first time that I have heard grassroots churches advocating absolute subservience by the wife to the husband. I am not sure why this theme is being plugged, but I have noticed that the recent appointment of the gay Anglican bishop in the USA is threatening to split the church with the African block leading the revolt, claiming that this could cause them to lose members to Islam. I wonder if the promoting of the above theme is just expediency.

After three hours, the service was over and the congregation left the church, having thoroughly enjoyed themselves. To me, it was all so different, having grown up with a mixture of Anglican and Calvinist where the church service was very formal and tight and not the place for dancing. I felt I had been to Mama Thembu's Wedding (Ipitombi). I think that because many people in Nigeria are not very affluent, and because of the irregular availability of electricity, the church supplies a large part of entertainment and a place for the masses to socialise.

Bridal couple and entourage

For those of you who have forgotten how we look

We unfortunately could not stay for the reception, as our time was limited, and we wanted to get to one of the local watering holes to watch the Bulls lose to the Cats (South African rugby/football teams).

I have spent some time thinking about Nigeria and the corruption and I most certainly don't understand all of it. Here are some thoughts and examples of it, and future letters will have more details and perhaps some insights.

Corruption in Nigeria is more than just bribes and back-handers. It is all over, and everybody, when they get a chance, will take it. We live in a hotel complex that has a laundry service. People who use the laundry sometimes find the local staff wearing their clothes. Besides the loss of clothing, people have the further dilemma of knowing that the clothes that are not pinched are considered by the locals to be ugly or out of fashion, and if none of your clothes ever get pinched, you really are square.

There is a mentality, which is ingrained, that when you are 'on top, milk it', and sometimes people have very little compassion for their fellow citizens.

Our maid works for us only three days a week, so when a new expat family arrived, we secured a position for her to work for them two days a week. After her first day there, she got paid, and promptly came to us to hand it over. The culture in Nigeria is if someone gets you work, the first month's salary and 10% of your earnings thereafter is theirs (obviously we did not take her money).

Travelling to our construction site, which is about 50 km outside Lagos, I sometimes have to go through seven roadblocks. Normally I get through without too much hassle, but if they do stop you, they are just after money. Normally you get a policeman with a big cheesy grin asking, 'What have you got for me?' Usually I try to act stupid (some might say that's not difficult) and look at him like, 'Why, what do I have of yours?' The problem is, taxi, bus and truck owners are hit the hardest, and if they don't cough up, they will find something unroadworthy on their vehicle and the vehicle will be impounded. The problem is that the vehicles are in an appalling state, and I believe that the biggest risk of getting injured in Nigeria is motor-vehicle accidents. I have explained to the locals it is not the vehicle owners who are paying but the commuter. In Nigeria it is no use making new laws and regulations because it is just another set of bribes that are extracted.

A further problem is that people don't get punished when caught; recently the police commissioner was caught with quite large amounts in his and his wife's bank accounts that could not be explained. He was simply 'retired' and no further action was taken. It is probable that he had enough information on all the others, so it was 'best let sleeping dogs lie'.

The photo is of a mansion close to where we live that is still under construction and has been for the last few years. There is continuous work being carried out, but no rush to finish. I believe that often the reason is that the building work is financed by bribe and extortion money, and its availability dictates progress.

Corruption in Nigeria is part of the culture and way of life, and I believe that all Nigerians know it's wrong, but greed presides. It is accepted very much like white South Africans eating too much red meat – they know it's wrong but still do it just because it is nice.

I will, in future letters, further address this issue. As an African, it troubles me, as it will keep Africa shackled to the poverty and mess it is in.

An interesting story, not from Nigeria but relevant to the topic. A colleague of mine had his laptop computer pinched in South Africa. Months later he received an e-mail from an Ethiopian, the proud new owner of his laptop (they hadn't even cleaned the data off before selling it at Johannesburg International Airport) seeking information about the computer. My colleague asked him to send some data files, which the Ethiopian did by cutting him a CD and posting it. (It's a miracle that the CD got by post from Ethiopia to South Africa.) Some months later the new owner again made contact, this time asking for the disc containing the drivers, as the laptop had lost some of its functionality. Clearly he thought they had a rapport. My colleague just ignored the request. I would have sent him a CD that would have permanently disabled the machine and then sent him an email saying it is possible this might have happened and when he is in South African again, he should look me up and I will rectify it. Should he have done this, I would have then taken my laptop back and very carefully explained to him where one should buy laptops, like at a shop and not from a thief at the airport.

I am off to Singapore tomorrow and should be back before Easter weekend. As you all know, Singapore is one of the most modern countries in the world (diametrically opposed to Nigeria) and e-mail works very well, so keep sending me your subscriptions.

Cheers from Nigeria.

Love and miss you all,

Arthur and Inez

Letter from Nigeria Twelve

Holiday in the East

April 2005

Since my last letter, I have hardly been in Lagos and have been to six countries on sixteen flights and have had both good and bad times.

As per the usual practice for Viva Methanol (the company I work for), travel arrangements are made in utmost secrecy and at the last possible moment. In January, I was going to travel to Estonia and was even telephoned and told to pack my things as 'you are leaving tonight', but due to various circumstances, I never left Nigeria. So when I was told that I was on standby for Singapore I took the attitude of let's wait and see. Arrangements changed back and forth, and finally on Thursday, it was confirmed I would fly on Friday afternoon. I received my tickets two hours before checking in. My flight was via Amsterdam, with an eighteen-hour stopover, and due to the time constraints, I had no visa (it would take at least a week to obtain in Nigeria).

My return date was not confirmed, as another company in the group was also interested in having me pay them a visit.

It took two-and-a-half hours in the Lagos traffic to the airport, only to find that, due to snow at Amsterdam, the flight had been cancelled. I was sent to the airline booking office on the mezzanine level. The office is about twenty square metres and was packed with about a hundred people, all trying to be rerouted, the bulk of whom had decided not to bathe until they got to Europe. It was chaos, with people pushing in the queue and pointless arguments about whether the flight should have been rerouted, etc. I eventually realised that unless things changed, all the flights out of Lagos would leave without me. I took charge, removed the 'pusher inners', closed the pointless arguments and we progressed. I got rerouted on Air France via Paris.

Rushing through check-in and immigration, I left for Paris Airport (Charles de Gaulle) with a twenty-one-hour stopover.

That week the news was full of the row between the Europeans and the Americans over subsidies to Airbus and Boeing, and I was surprised that Air France was using a brand-new Boeing 777 (a very nice aircraft indeed). I imagined the American negotiator telling the French, 'What is wrong with Boeing? Even Air France use it.'

We arrived on schedule at Charles de Gaulle, but as usual for flights from Nigeria, the plane was parked in the far, distant corner of the airport and the passengers bussed in small groups to the terminal. At the terminal we were met by the police and each individual's passport was thoroughly examined and a number of people 'removed'.

I used this opportunity to ask the policeman if I could get a day visa so as to spend the day in Paris. His response was, 'No problem, provided Air France agrees,' and so my adventure with French bureaucracy began.

Having spent most my working life with Johannesburg City Council and Eskom (electricity utility of SA), I have learnt the following about bureaucrats:

- They don't like to think and reason and then make a decision.

- Apply the rules, because they are the rules irrespective of any other factors.

- Make sure no one else makes decisions (i.e., blindly apply the rules) when it's your job.

- Never apply common sense.

First I went to Air France to get their permission, but at the counter I was told that as my next flight was leaving from Terminal 2B, I had to go there for permission (I was at Terminal 2F.) No problem, through security, catch the bus, more security and on to the Air France counter. Stood in a long queue, eventually got to the front and explained my story and the response was no, asked why and they said the police wouldn't allow it. First I tried to explain, then argued and finally begged, but no luck.

After two sets of security and a bus ride I was back at Terminal 2F, and I went to the Air France counter with the story that the police had agreed to a visa; I just needed Air France to give a note. After long deliberations in the back office, a lovely mademoiselle eventually accompanied me to the airport police station.

My story was explained, and the officer behind the counter agreed and took my passport for stamping, but just then the commanding officer walked in, shat himself – this was his decision to make – and then said no visa. I still tried to reason with him, but he left. So close but yet so far.

I decided to take one last stab at it and tried to go through immigration, but no luck – no visa, no entry. I was beginning to miss Lagos, as here it would have been easy to get a visa – just pay someone.

I resigned myself to a long day at the airport and decided to get some breakfast. After two sets of security and a bus ride to Terminal 2B, I only found one open restaurant, with almost nothing to eat. Someone said to me that Terminal 2F had better restaurants, so I decided as I had all day, let's go there. This time they wouldn't let me through security, as the rules say I must stay at the terminal that I am going to depart from. I argued (really argued) that this was irrelevant and that there was no food to eat here, and when I said I didn't consider a stale bun (croissant) to be food, the security staff adopted the tactic of ignoring me. I noticed there was a door that was used by the staff to bypass the security point, and when no one was looking, I nipped through and caught the bus to Terminal 2F.

Now the same problem here, I couldn't go through, as my flight did not leave from here. I argued like hell, but the security got aggressive and threatened to call the police, so I left but decided to go to them anyway and insist on permission to go into Terminal 2F; after all, I had all day to entertain myself.

At the police station, a new shift had come on duty, so I started to explain my story from scratch to the officer at the desk, and before I got to the part about the food and that Charles de

Gaulle Airport was as bad as Lagos (this is true), he said to me I should be able to get a visa, but as they were busy starting a new shift, I must come back in forty minutes.

When I returned this officer was missing, and a new one was behind the counter. I decided it was best if I helped with the decision-making process and told him my visa had been agreed to by the police and Air France and I had just come for the stamp in my passport. OK, he said, and took my passport to the back and eventually returned and said they couldn't issue me a visa, as I was not rerouted. No, I said, I was supposed to fly via Amsterdam. He asked me where my visa for Amsterdam was. I replied, 'It is not required, and they always allow me into Holland when transferring flights' (I was taking a chance here, but figured he could either accept it or reject it). 'OK,' he said and disappeared to the back. About fifteen minutes later, a nineteen-year-old brat of a policeman turned up with my passport and proceeded to lecture me on how lucky I was to get a visa and it was free, so I had to spend my money wisely (I think he was hinting at a gift). During all this, he had my passport in his hand and kept waving it in my face. When he eventually finished, I grabbed my passport and said, 'I hope you beat the English at the rugby this afternoon,' and hot-footed it out of there through immigration and into the first taxi to Paris.

An hour later, I was dropped off in Paris and there were six inches of snow on the ground and I was freezing and wondering if it was all worth it. However, I spent a wonderful Saturday walking around Paris, and when I returned to the airport, the immigration clerk was a bit dopey and could not find my visa, so would not let me in. I was tempted to play along, but as my flight was due to leave, I helped her out.

I have the view that in order to become an immigration official, being rude and arrogant and stupid is a prerequisite. Furthermore, why do we have to fill out immigration forms in triplicate with information that is in your passport that is barcoded, and the information is scanned in anyway. All it proves is that I passed my grade-two comprehension test.

Next stop was Singapore, with Singapore being Singapore – that is, clean, over-organised and expensive.

Below is a picture of a platform on the subway (the doors open and you step onto the train). Getting on a train here feels like stepping into a fridge, and there is none of the charm of the London Underground, with its buskers and bright posters, etc.

Once the dates of my business meetings were confirmed, I realised that by taking two days' leave prior to the Easter weekend, Inez and I could spend Easter in Bali. So there were some rushed bookings and arrangements, and two weeks later, Inez joined me in Singapore.

We were given tickets to the Malaysian Grand Prix, so the first weekend we flew up to Kuala Lumpur for the race – the jet set or what? This next photograph is of our view of the track.

I took the picture below of this the 'modern Muslim lady' and one would think the natural technical progression would be to give her windscreen wipers (and if you put hair on them, they could double up as artificial eyelashes). But sadly, a technical advance will probably be to fit a remote-controlled closing device, so her husband can control what she sees.

After the Grand Prix, we returned to Singapore, and Inez stayed with her cousin Annabelle and I went to Indonesia for a short business trip. Once again, a big thank you to Annabelle, who helped make the visit enjoyable and gave us free accommodation.

Later in the week, we flew to Bali for the Easter weekend. Although Bali is part of Indonesia, it is a separate island and very different from anything I have seen or experienced. Besides being a South Pacific island that is cheap, has wonderful beaches and great hotels, it contains a wealth of culture (all new to me) and magnificent scenery, which includes a semi-dormant volcano.

In the photographs below are pictures of terraced rice paddies and the volcano.

Terraced rice paddies

The black areas in the middle ground of the next photograph, are the lava from the last eruption in 1994.

Although Indonesia is Muslim, Bali is very much Hindu, and life revolves around it, and everywhere you go, you experience it. My knowledge of Hindu is limited, but it is my understanding that the rituals and ceremonies have changed a bit from the more traditional practices we have been exposed to. Every house has a temple, and offerings of fruit, biscuits and cigarettes are continuously made to the gods.

All the temples have statues of monsters 'guarding' them, and as there are temples everywhere, Bali is a bit like a scene from a horror movie, especially at night. We arrived at Bali quite late in the evening, and Inez immediately went to explore the hotel (I think it was to find coffee and cake) but returned soon, a bit spooked, as all the open areas are full of these statues.

The gateway above is known as the split gate, and legend has it that if you go through and you are evil, the two halves will close and crush you. I, of course, had no problem, but some of you might have to think twice before entering.

And sometimes life is just great!

Pool bar

I can recommend Bali as a good holiday destination.

Inez and I returned to Lagos at the end of March, and ten days later, I was on my way to Estonia via London.

Estonia was part of the former Soviet Union and is the most miserable and saddest place in the world. It is cold, dull and full of grumpy people. The Estonians make it clear that they are of Nordic descent and hate the Russians. About a quarter of the population are Russians (they moved there during the Soviet era) but cannot get Estonian citizenship unless they pass a stiff Estonian language examination. We all thought apartheid was bad.

Below is a typical view of an Estonian street – drab, with neglected buildings. One evening I sat in a pub and besides orders for drinks, no one said a word for more than an hour.

Shopping is the exact opposite to Lekki Market in Lagos; here the shop assistants ignore you, and if you do ask for help, they are rude. Although they have had no communism for more than a decade, its influence is still prevalent – there is little advertising or marketing and few bright colours.

Tallinn

The capital of Estonia is the city of Tallinn, which has its origins as a medieval city. The old town has been restored and is a 'waterfront'. I found an Irish pub and thought I had found a place with a warm atmosphere, but alas, it was filled with a few locals, all keeping to themselves. When a touring Scottish rugby team arrived, half-drunk and wearing kilts, all the locals got up and left. I think they knew they were going to find out first-hand what the Scots don't wear under their kilts.

Estonia is now part of the European community but is still somewhat cheaper than the rest of Europe. In Sweden and Finland, alcohol is heavily taxed, so many go to Estonia to drink/catch up. On the weekend, you see them sitting all day in the bars, rectifying their blood alcohol levels.

When I returned home, my mother and mother-in-law had arrived in Lagos for a visit. I was able to get two days off work, and a convenient public holiday gave me some time with them. We showed them most of what Lagos has to offer. We took them to a friend's beach house for a barbeque. In the picture below is my office cleaner picking coconuts. I bet there are not many of you whose cleaner has these skills.

Life is slowly returning to normal; however, the rainy season has begun, and this week the fuel tanker operators have gone on strike, so the Lagos traffic is slow and tedious.

I am gradually catching up with my letter-writing and I apologise to those whom I still owe letters; however, subscriptions are still required for 'Letters from Nigeria', and many owe me for the last two letters.

Love and miss you all,

Arthur and Inez

Letter from Nigeria Twelve A

X-Rated Version

April 2005

Hi,

I have been writing these letters from Nigeria for a year now and, as they have a fairly broad distribution, have tried to keep them reasonably respectable. It is difficult, as many times human nature is at its best when the carnal needs and desires are inspiration. I felt I had to share these 'I wish you were here' moments.

I don't like Singapore, as it is very clean, sterile, and super-clinical and the people are naïve, bordering on American ignorance on matters outside the USA. The laws are strict, and if you fart loudly in public a cop comes over to sniff, he might decide to lock you up.

So imagine my surprise the next evening when I was in a taxi and the driver said, 'Need a girl?' (I was in the area Little India to do some shopping; it's where a big department store and the computer shops are). Being a good traveller and wanting to make full use of all the opportunities I get to learn about the locals and their culture, I said, 'Where?' I thought I could also finally see for myself if it was true that a Chinese girl's fanny is horizontal.

He then proceeded to tell me he could take me to an area where there were karaoke bars and the girls take everything off. OK, I said, let's give it a try, it is still early. I know the Chinese are fond of karaoke, but the name is probably just to front it. Ten minutes later, he drops me off in the seedier side of town, and I proceed to the first karaoke pub I see.

When entering such an establishment, I expected to find a girl dancing around a pole, so imagine my surprise to find a punter with a mike in his hand squealing out the songs of the Everly

Brothers, and the girls sitting and being the fan club. The girls must have been thinking, *We have a shit job and have probably had to do some bad stuff, but listening to this singing must be close to the worst.*

I decided to move on and went on to a normal pub only to find it full of prostitutes of all nationalities, but mostly Oriental. They would approach me very directly in their broken English – you stand there and she walks up, no hello, kiss my arse, nothing. Just, 'You want to make fucky?' The beer was more than R60.00 a glass, so I did not stay long and left.

The next evening I caught a taxi and was on my way to meet with Inez's cousin when again the taxi driver made himself available to assist me with my sex life. The conversation went some thing like this. Text in italics are my comments or explanations.

TAXI DRIVER (TD):   Where you come from?

ME (M):   South Africa.

TD: Huh, where?

M: SOUTH AFRICA!

TD: You like Singapore?

M: No, too expensive.

TD: Singapore good, got everything.

M: Really.

TD: You want girl for fucking?

M: No, too expensive.

I began to realise I was going to get no peace with this idiot.

TD: You get man-girl.

M: Huh?

TD: Verlee good.

M: Fuck China, a transvestite sissy.

TD:	No, verlee good, you fuck her, she fuck you, all for $200. Aaah verleeee good.

You have to hand it to him, this was a two-in-one marketing package.

TD:	You want to go now?

He was determined.

M:	Sorry, I must meet a lady now.

TD:	No problem, you bring her with, man-boy fuck you both.

M:	No, umm, we are going to have food now, no time.

TD:	Here my number, you phone me, I come and take you.

I was not winning.

M:	I can see the place where I am meeting my friend, here is money for the fare, keep the change, bye.

I left the car in a rush, went into a shop, waited ten minutes, and came out to catch another cab to take me to my destination.

So, after all that, I still do not know the orientation of a Chinese girl's fanny.

I will be in Singapore till the end of the month. Inez joins me next week and we will spend some time in Bali.

Cheers,

Arthur

Letter from Nigeria Thirteen

General

May 2005

The last letter was drafted in a bit of a rush, as it had been two months since I had written and I was somewhat busy at the time. I forgot about a few items, which will make up the first part of this letter. Also, Inez has put pen to paper and written some articles on the funny side of life in Lagos, which I have also included.

First and foremost, it has now been more than a year that I have not smoked – in fact we are going for fourteen months. How many of you thought that was possible?

In my last letter from Nigeria, I touched on my trip to Estonia. Although I found it a very sad place, I was intrigued by the effect of fifty years of communism. I have always admired the Americans for their work ethic, high levels of service, independence and energy. Estonia is so different. The background to my visit is as follows:

- The paper factory was built in the 1930s.

- Some modernisation took place in the 1970s, all Soviet equipment (very old-fashioned).

- With the fall of communism, the factory was privatised and went bankrupt in about eighteen months.

- The group I work for bought it as scrap and added some new equipment and manage to run the plant at a small profit.

- Gas is required to make steam for the process and comes from Russia.

- Gas prices are going up and the plant is no longer profitable.

- Arthur is to go to the plant, perform some magic and halve the amount of gas that is burnt.

Prior to my arrival, an accountant from Singapore had reviewed the costs, and by doing some accounting magic, showed that big savings could be achieved. Unfortunately, he was also trying to change the laws of physics, and I spent quite some time teaching him thermodynamics. Although he finally accepted some of the principles, he was still convinced that if I looked at it from a 'financial perspective', miracles were possible. I have now changed my view of accountants from being little grey men to very creative people.

Dealing with the people on the plant was difficult. First there was the language barrier, and then I was treated with deep suspicion, as most people from the head office are. After I had spent some time in the plant, it was evident that a lot of energy was being wasted, and the reported figures were all rubbish and everyone was super-defensive about them. To put this in perspective, one must understand the culture from the Soviet times. All planning was done centrally, and this factory made paper bags for a cement factory (nothing else). Provided the quota number of bags were made, there was no problem. Gas, electricity and raw materials all came with no charge, and figures like efficiency and productivity were manipulated to keep the 'powers that be' happy. If you wanted something, you put in a motivation to Moscow and it might eventually come; if not, you're covered, as you've done your bit.

I hardly believe that I am qualified to state the reasons for the failure of communism as an economic model, but this example speaks volumes.

Returning to Lagos. There have long been rumours of Nigerians who have hyenas and baboons as pets, which they parade around the streets of Lagos as a type of circus trick. Our mothers and Inez were 'fortunate' enough to witness this exhibition. Looking carefully at the photograph, you will note that the hyenas, besides being well known for having the strongest set of jaws, are also very big, strong animals, and although I would definitely avoid getting into a fight with the handler, if the hyena decided that any paying or non-paying spectator was lunch, it is probable that the hyena would not go hungry that day.

I have some more photographs that deal with transport in Nigeria. In Nigeria, like South Africa, the most dangerous activity you can undertake is to drive on the roads (Nigeria probably being somewhat more dangerous). Generally the vehicles are dilapidated and the drivers are ruthless, with no law enforcement.

In the photograph below, the truck was hired by a friend to transport building materials, but when he arrived at his destination, that truck was so overloaded it was not able to mount the kerb. No problem for the idiot of a driver; he promptly reversed the truck and came charging forward, hitting the kerb at some speed, dislodging the load and changing the centre of gravity, resulting in Lagos's one and only ski ramp (mobile).

Below is how vehicles die in Lagos. They get driven until they are dead and then just abandoned and left to slowly rot away. The signboard was annexed from a nearby property development and placed there to warn oncoming traffic of the stationary vehicle, but it does have a touch of irony.

Yes, it is a horse inside, and the only reason why the back door is open is because the bench does not fit anymore. Note there are also other passengers inside the taxi. I wonder what fare the horse was charged and if eventually one of the passengers will complete his journey on the back of his fellow passenger.

In the photograph below, there is another demonstration of the African belief that when you load a vehicle and you overcome the boundary of the roof, the sky is the limit. Examine it carefully and note the additional baggage and the messages.

In Nigeria, religion is big business, and there is good streetwise business acumen. This billboard has recently appeared. The reference to the Bible refers to when Jesus delivered the parable from a boat on the sea of Galilee about the farmer sowing seeds, and when, afterwards, a storm brewed and the sea became turbulent and Jesus commanded the wind and the sea to be calm. I leave you to draw your own conclusions. This letter was written shortly after the Boxing Day Tsunami of 2004.

Billboard

Holy water?

ARTICLE ONE: CUSTOMS CLEARANCE

Earlier this year, in preparation for a trip to Singapore, I bought four metal sculptures depicting a group of musicians playing traditional Nigerian instruments. I intended to give them away as gifts in Singapore. They are very beautiful but are not antique, nor are they unique, as the sculptures are made in volumes by the locals and sold at local markets.

After carefully wrapping the gifts in protective paper and gift wrap, I packed them neatly in my suitcase and took off to the airport. At check-in it is necessary to have your luggage checked by customs to ensure that nothing illegal is taken out, or at least not without a bribe. The customs official inspecting my luggage found the beautifully wrapped parcels and enquired what they were. When I explained that they were metal statues purchased at Lekki Market, he started violently tearing at the gift wrap and behaved as though I were an art thief. He demanded the necessary papers to export 'Nigerian antiquities' and accused me of 'breaking the laws of our land'. I realised that this guy was determined to extort money – otherwise he would continue to damage my beautiful gift wrap – and quick action was required. I dived into my handbag and pulled out my make-up case – his eyes lit up, thinking I had pulled out my wallet, and he suspended the destruction of the gift wrap, only to see me pull out a handkerchief, which I used to wipe the pseudo-tears from my eyes. Seeing that I was not going to give him any money – clearly I was below minimum IQ or lacked Nigerian savvy – he totally despaired and roughly threw the 'antiquities' into my bag and told me to go. I shed tears of joy as I left, because once again I had avoided paying a bribe.

Markets like Lekki exist almost solely on tourists, and if tourists are harassed and bullied into paying bribes every time they buy curios, they will stop doing so. The sad part is that the poor artists who eke out an existence from these curios will suffer the most.

ARTICLE TWO: BEARD MAINTENANCE IN BALI

By the time we reached Bali, Arthur had been away from home for some three weeks. In that time his beard had grown quite scraggly and was in desperate need of a trim.

On a walk around through the streets of Bali, we passed by a little barber shop, and the young Balinese barber was in attendance. I suggested to an always-sceptical Arthur that he pop in and ask the barber for a trim.

Bravely he walked up to the eagerly awaiting barber and asked him, 'Do you speak English?' to which the barber wildly nodded his head and said yes. Then Arthur, gesticulating towards his chin, asked the barber if he would trim his beard just slightly, just to make it neat and tidy, not a total shave, could he do this? Again the barber wildly nodded his head and said yes. Growing in confidence and seeing the same razor as he used at home, he instructed which blade and cutting position must be used and again reiterated, 'Not too short.' 'Yes' smiled the barber, still wildly nodding his head. With his mind at ease, Arthur sat down in the barber's chair while the barber, now also relaxing, grabbed his razor and dived into the task of shaving his beard. Moments later, Arthur sported a bald patch on his cheek. Realising that the conversation held in English had been decidedly one-way, Arthur jumped up, wrestled the razor from the young man's grasp and finished the job himself, paid the man and left in a huff.

The bald patch did not look so bad and, after all, the mission was accomplished. The barber, on the other hand, felt overwhelmed, but could really not complain, as he got paid for a job he didn't do and learned a few useful English and French words in the process.

ARTICLE THREE: TIME'S ARROW

One of the biggest problems we have living in Nigeria is with finding replacements or spares for the things we brought with us that, through the rigors of normal life, get damaged, lost or broken. Just such a thing happened to a friend of ours when he noticed that his watch had slowed down. It was a beautiful, good-quality watch. The possibility of obtaining a new battery for his timeless timepiece in Lagos was well nigh impossible, so he had no option but to replace the watch.

This was an easy thing to do, as there are numerous street vendors who will sell you anything from a packet of sweets to magazines to stationery to shoelaces to sunglasses to watches and more, all through your car window (this redefines window shopping; here you shop from your car window). Finally he purchased, for US $12, a handsome wristwatch with a brown leather strap. He was particularly proud of himself, as negotiations started at US $120.

Wearing his new watch proudly, he was pleased to note that for the first two days, it kept accurate time. However, on the morning of the third day, he noticed it was losing time, and by that afternoon, it had slowed down to a complete stop. Refusing to believe that his new watch was already broken, he took it off his wrist to reset the time. That is when he felt it click and it tweaked some faint memory still lingering from his childhood and he realised that his new watch was not battery-operated but a wind-up.

ARTICLE FOUR: GENUINE LEATHER

A South African friend who has plans to be here for a few years decided to purchase a new lounge suite.

She went to see a furniture manufacturer who was very helpful and showed her several samples of genuine leather so that she could choose the exact colour. As she had already lived here for some years, she was brutally aware that Nigerians will always try to scam you at every possible opportunity. She was determined not to get caught, and during the negotiations, she stressed the importance of the lounge suite being made from 'genuine leather'. Yes, madam, we will use genuine leather.

Three weeks later, when it was delivered, the lounge suite was in accordance with the agreed design and colour but was obviously not leather. So she asked him, 'Is this genuine leather?' 'Yes' came the reply. With disbelief she asked, 'Where did it come from?' and he answered, 'It is genuine leather made in the factory.' Exasperated, she exclaimed, 'Leather is not made in a factory, it comes from the skins of cows.' 'Aha, madam, you don't want genuine leather, you want real animal skin. That is something quite different.'

Thanks to all our friends here who assisted with contributions and photographs.

Miss you guys.

Remember, subscriptions are due again, and they do help keep us sane, or at least as sane as you remember us.

Cheers,

Arthur and Inez

Letter from Nigeria Fourteen

Living in Lagos

June 2005

Another month has passed, and all is quiet in Lagos, Nigeria.

This letter consists of a lot of photographs, and it is not because I regard my audience to be illiterate, but rather that a picture paints a thousand words.

There have been reports in the international press of trouble in Nigeria, and much like the good old 1980s and 90s in South Africa, you have to watch the news in the evenings to find out about all the bad things that went on around you during the day.

I have noticed bad/alarmist reports tend to surface with a sudden increase in the oil price. The latest incident relates to a scare that a car bomb would be placed outside a foreign consulate in Lagos, more specifically that of the United States. The northern part of Nigeria is largely Muslim, and there appears to be Al-Qaeda infiltration/influence/sympathy. However, because of the rude and arrogant manner in which my colleagues and I are treated by foreign consulates each time we apply for a visa, we could also become suspects.

The biggest disruption to life caused by the bomb scare was that access to the road in which all the embassies and consulates in Lagos are located was via a roadblock. This is the only relevant and legitimate roadblock in Lagos (which makes a change from routine extortion points) but causes lengthy delays in getting access to the embassies, and needless to say, it is just the time when I need to get a Schengen visa. More importantly, in this road is one of the better watering holes in Lagos and the venue of the only live rock band, and access is also severely restricted.

On the social front, Inez's 41st birthday was celebrated with a fine dinner party at our house.

A friend of ours from the German community invited us to a major shindig at his house and yes, we did eat the fellow below.

With the rain comes the fly menace. When you are outside, the moment you bring out anything to eat or drink, they descend upon you, rather like, well, flies. This covered beer glass is standard-issue at the bar to prevent you having to share your drink with the flies, or drinking the flies with your beer.

About an hour and a half's drive inland from Lagos is Shagama Golf Course. The topography changes dramatically, with hills and valleys, but the tropical vegetation remains. The golf course is surrounded by walls, and when inside, you feel as though you are in a different world and it reminds me of the Lowveld in South Africa. There are many Chinese expats in Nigeria and they all play golf here; hence I refer to it as Shanghai Golf Course.

We are in the middle of the rainy season, and although it is the middle of summer, temperatures are cooler and much more pleasant. The rainy season in Lagos is characterised by:

- Air conditioners cope and we now have to take them off maximum.

- Daily power failures.

- Daily loss of DSTV.

- Regular breaks in internet access.

- Massive thunder showers.

- Endless traffic jams.

The project I am here for has picked up, and in the course of my work, I have been driving around Lagos to attend meetings with various parties. I have taken my camera along and have been snapping pictures of suburban Lagos through the car window en route.

Suburban Lagos

Lagos city centre

Nigerian National Theatre

The magnificent structure of the Nigerian National Theatre, which is hardly ever used, is just another way of wasting money as there are far greater needs and there are also better ways of spending money on the arts.

The architecture in Lagos continues to amaze and confuse. The area we live in is very wealthy, and having pillars on your house is 'the in thing'. The architect who designed this house was either confused or extremely forgetful and forgot the bottom half of the pillars.

This building contractor is the only honest one in Nigeria.

The guy below can give you a comprehensive service – no matter what the outcome, he still earns his fee.

As mentioned in the last letters, our moms came to visit. Who says you can't have fun in Lagos?

Cheers!

I should be off to Germany at the end of the week for a week or two.

Love and miss you all,

Arthur

Letter from Nigeria Fifteen

Lagos and Tallinn

July 2005

Since my last letter, I have spent more than half my time out of Lagos in Germany, Estonia and Holland.

Germany remains Germany and is clean, wealthy, sophisticated, over-organised, and the Germans are just friendly enough so that you can get by.

Estonia was in the middle of summer, and the problem is the sun won't go down in the evening, and when it eventually does, it is hardly down and then it comes up again, I still have a bit of animal instinct and find it impossible to go to sleep when the sun is up; on the other hand, I think the sunshine helped to cheer up the people a little. An example of their behaviour is when an Englishman and I were sitting alone in a bar and the music was so loud any conversation was impossible. When we asked the barmaid to turn down the loud rap music, she aggressively demanded to know what was wrong with it. When the Englishman tried to explain that music must be melodious and I added that rap is not music but just a bunch of overweight African-Americans complaining to a slow beat, it resulted in her thoroughly scolding us and telling us that this music was appropriate and if we didn't like it, we could leave.

The barmaid's attitude is, I suspect, still a bit of a leftover from Soviet days, where the concept of the customer and good relations was not there and the bar was being run as a state-owned monopoly (e.g., the post office).

I also found many of the marketing skills that we take for granted missing, like when going through a menu, the products sell themselves; however, reading a menu in Estonia is like studying a software manual. You can never find anything you want, and if you do find it, you can't understand it, and when you think you have mastered it, you still don't get what you wanted.

The Estonians are good-looking people with some of the most beautiful women I have seen, but they are all so sad. If you do meet a friendly happy person, it is generally an overweight, tattooed British tourist (both men and women).

Estonia is a country of contrast, with some beautiful buildings and areas and on the other hand some very bland, dull and run-down areas. I was able to get some time off and spent some time getting around the city of Tallinn, and the photos below show the scenery and some of the contrasts.

Panoramic view of the Old Town

The Old Town Square is very much an entertainment centre for locals and tourists.

Downtown Tallinn has a combination of modern and old buildings and there are modern shopping malls with good shops and with prices very much in line with South Africa.

Downtown Tallinn

The building on the left is the hotel where I stayed on the fifteenth floor.

View across Tallinn

Ship in Tallinn Harbour

One day I had decided to take a walk to the harbour to look at the ships (in the summer, Tallinn is a favourite port of call for many ocean liners). But after walking for about two kilometres, I realised that I was lost and asked for directions. Again, the old Soviet Union mentality came into play, and as foreigners are treated with suspicion, I struggled to get people to share information, and eventually I found an ice-cream vendor. After I bought both of us an ice cream, he opened up and agreed to help. He was adamant that the harbour was at least 2–3 km away and that it was too far to walk and that I should catch the No. 15 bus. I pointed out it was Sunday and would there be buses? No problem, at least one every ten minutes, very much part of the old communist Eastern European era, excellent and cheap public transport.

As luck would have it, although I did catch the No. 15 bus, it was in the wrong direction, and I had a non-tourist tour of Tallinn, and here lies the contrast. There are many dilapidated buildings and a skyline that is populated with pre-fabricated concrete apartment buildings.

In the next photo is a sight often seen, and although it appears that the house is abandoned, it is not – people actually live there.

'Abandoned' house

The photo below is interesting from a town-planning point of view. Residential areas were integrated into industrial areas, and the worker lives close to the factory gate. Again, all so different from what I am used to.

Houses next to factory gate

I met an Estonian in the pub and had an opportunity to chat with him, and I asked him if he was better off now than he was in Soviet times. He replied that they were better off in Soviet times, as although they had shops with no goods to sell, they had money. Now they have shops and can buy anything but have no money and no security. Previously, come what may, you would always have food (maybe bland) and housing (maybe a compartment in a concrete tower). My view is that it was easy, convenient and there was no pressure or responsibility. Perhaps the inverse is the formula of success of the wealthy Western nations.

On the way back, I had a night in Amsterdam, and saw my favourite aunt, Cecile, and her friend Yvonne. Thanks to them both for a great night out in Amsterdam.

While in Germany, I had to stock up on vitamins, and I went into a health shop for the purpose. The shop assistant was very helpful but was a bit concerned about the quantity I was buying. I explained that I live in Nigeria and that I was stocking up. She was intrigued about Nigeria and pointed out to me that I was lucky, as I had access to organic fruits and vegetables, meaning that fruits and vegetables are grown:

- in real soil with no artificial matrix material;
- no fertilizers are used;
- no insecticides are used;
- there is no genetic modification;
- no radiation to preserve.

Yes, sure, we do buy our organic fruits and vegetables from our vendor, Patience, at Lekki Market, which in itself is an experience and, after all, going to Woolworths[10] is boring and just too easy.

---

[10] A chain of upmarket grocery stores in South Africa.

Fruit at Lekki Market

Let us examine what organic fruits and vegetables means:

- Grown in real soil, half of it comes with.

- I am sure fertilisers are not directly added, but there is bound to be adequate waste products in the vicinity of the fields, which suffice adequately.

- The insect kingdom has normally had their share prior to purchasing.

- Genetically modified grapefruit are round with smooth skins; ours come in a variety of odd shapes with thick skins and an abundance of pips.

- The lack of any preservation measures means that each item has at least one spoilt part.

We are still in the middle of summer, which for Lagos means almost continuous rain but far more manageable temperatures. On one day we measured 1,450 mm of rain. The picture below shows the flooding at our construction site.

In the picture below is a technique most often used here to pour concrete slabs. Although it is a great way of creating employment, I could think of no worse job than walking up and down the scaffolding all day with a heavy load of concrete on my head.

Again, Inez has put pen to paper and produced another article. Should all the subscribers make their payments, they can also use the opportunity to encourage her to do more of these fun insights into life in Lagos.

ARTICLE FIVE: NIGERIA – FULL OF SURPRISES

A few days ago, my friend Charlotte and I went to Lekki Market to buy her a new pair of shoes.

Inside one of the stalls, we were discussing the wares in Afrikaans and the saleslady, a local Nigerian, hearing our foreign tongue, started speaking to us in German. This gave rise to a lengthy discussion where she explained to us (in English) that she had strong links to Germany, she had even lived there for a while and had sent her children there to study. Furthermore, she buys many of her products there.

Charlotte purchased a totally amazing, over-the-top, wild pair of pink stilettos. Our Germanic Nigerian claimed they were imported, but soon the realities of Nigeria returned, as generally Nigerian manufacturers of fake goods have remarkably poor spelling abilities, and printed inside the shoes in gold letters is 'Mode in Italy'.

I am reminded of the story of the man who asks the Pope, 'I would like to come to Rome for a week, what do you think?' and the Pope replies, 'A week would be good.' 'What,' asks the man, 'would you say if I wanted to come for two weeks?' and the Pope replies, 'Two weeks would be even better.' Again the man asks, 'What if I wanted to come to Rome for two years?' 'That,' says the Pope, 'would not be long enough.' That is how I feel about Nigeria. I have been here for over a year and every day I get a new surprise. That is the thing about Nigeria, first you learn not to expect anything and then you get surprised by what they can deliver.

Inez and I have been well, we are planning to spend some time in Europe and Holland in the second half of October, more details to be given as the picture becomes clearer.

Love and miss you all.

Cheers,

Arthur

Letter from Nigeria Sixteen

Partying

August 2005

As promised, this edition will deal with entertainment, or at least the best that we can do from an expat's point of view. There are many things to do, such as nature hikes, golf, yachting, etc., and I have previously dealt with some of them and now will only deal with partying.

There are many restaurants, with all the major varieties represented, e.g., Chinese, Thai, Japanese, Italian, French, Mexican (at least from an American point of view) and even a poor copy of a Planet Hollywood. Generally they are all mediocre and rather expensive and invariably owned by Lebanese and have a Mediterranean flavour.

There are many nightclubs/bars frequented by expats with varying degrees of sleaze. One of the better-known ones is Bob's Bar, seen in the photograph below.

Bob's Bar can at best be described as a shack on the waterfront, with most of the patrons sitting outside in the car park on plastic garden furniture.

It is owned by an ex-Royal Marine (Bob) who has settled in Lagos. The main attractions of Bob's Bar are a live rock 'n' roll band on Friday nights and the lovely local Lagos ladies of ill repute. They are referred to as mosquitoes, as they are forever buzzing around, making a nuisance of themselves with their aggressive marketing techniques, and they also *steek*/sting. However, when you are accompanied by your wife or girlfriend, they leave you alone, and Inez now refers to herself as mosquito repellent.

Rock 'n' roll band

Bob on the drums, with his traditional hat and loud pants

Who says geriatrics can't rock 'n' roll?

The band is made up of expats of various ages and abilities, but as a rule, they play good rock and roll. We have had some good nights there, rocking till the wee hours.

Nearby where we live is Alpha Beach, which is a hangout for locals and good for after-work drinks. Alpha Beach is very typically Lagos; i.e., it is noisy, dirty, chaotic, crowded but has lots of energy. The white sandy beaches are lined with grass-roofed shacks where you can get drinks at good prices and sample the local cuisine, and all with extremely loud soul music blaring in your ears.

Alpha Beach

In the photo below, Inez and her good friend Charlotte are wading in the warm Atlantic Ocean.

As at most seasides, pony rides are on offer. In the picture below, the lady (who is actually from the east end of London visiting relatives in Lagos) was screaming at the top of her voice whilst riding the pony; however, I thought the poor pony had a lot more reason to scream.

Soccer is very popular here, and below is an official sports ground with a club match under way. Note, not all the players have the correct kit, and spectators use the surrounding highways as grandstands.

As many of you know, Inez's sport is walking; however, her favourite is to combine it with shopping, and if ever you go overseas with her, you learn that most time is spent walking and shopping, normally with me doing my best to keep up with the pace of both the walk and the bills while carrying the bags. I have previously dealt with shopping in Lagos, but as it is entertainment for Inez, I will touch on it again.

We recently found the Chinese market, where you can get a variety of goods at good prices but somehow one always feels the word 'disposable' is missing from every label. The relevance of the theme of a medieval castle in the next photograph is beyond me.

Chinese market

At the toilet inside the market we found this sign.

I am sure you are asking the obvious question: How do they know which undignified act of humankind took place?

Many subscribers' subs are way behind, and I have decided to generate a renewed interest by running a competition with fantastic prizes. All subscribers need to do is send us an e-mail with solutions to the above question. For one knows that 90% of you cannot resist a chance of winning a prize.

There is a modern Metro Cinema in Lagos that shows most of the current 'action' movies on circuit.

Metro Cinema

The shopping complex is very Western and has a bookstore which looks similar to Exclusive Books. We spent some time there and I noticed that the books were divided into four sections, namely: children, fiction, religion and everything else. In the fiction section I found the dictionaries, and I am not surprised when we see billboards and signs like these below.

The plaza is actually UNIQUE PLAZA, and one would have thought that a dictionary would have been consulted before going through the expense of installing the large neon signs. The other is a good example of pidgin English; the meaning is obvious, although it displays a curious logic.

There is no rush in Africa

On a tragic note, the practice in Lagos is to place your house, shop or business as close as possible to the road, if not in the road, as can be seen in the photograph, I have seen this practice elsewhere in Africa too.

Shop placed in the road

Two weeks ago outside our compound, a truck drove into a restaurant located right next to the road, killing eleven diners. The reaction of the public was to burn the truck. The sad thing is that the main causes (i.e., unroadworthy vehicles and the building of shacks immediately adjacent to the road) are not dealt with; however, the endless roadblocks by the various branches of the police checking for 'unroadworthy vehicles' and enriching themselves continues.

URGENT CALL TO PROMOTE GOOD GOVERNANCE IN NIGERIA!

WE IMPLORE ALL HEADS OF CHRISTIAN CHURCHES, MOSQUES & INDIGENOUS RELIGIOUS ORGANIZATIONS TO PREACH AT LEAST 5 SERMONS AGAINST CORRUPTION BEFORE THE END OF THIS YEAR TO PROMOTE GOOD GOVERNANCE IN NIGERIA.

THIS MESSAGE IS FROM:

GOOD GOVERNANCE SOCIETY OF NIGERIA,
CENTRE FOR HUMAN DEVELOPMENT,
14 BOLA BABALAKIN CRESCENT,
IKOLABA, AGODI GRA, IBADAN.

This is an ad I have taken from page three of one of the large newspapers, and if it will have an effect I don't know, but at least some effort is made, even if it is by a society and not the government.

Life in Lagos is still pretty much the same: the rainy season continues and the temperatures are moderate. Remember, answers are required with your subscriptions.

Love and miss you all.

Cheers,

Arthur and Inez

Letter from Nigeria Seventeen

The Leaving of Lagos

September 2005

Inez and I are packing up, and we will leave Nigeria permanently at the end of this month. I have lost confidence in the project and lately get the impression that the owners are beginning to look at alternative development options for the site.

I joined expecting to get involved in a large billion-dollar project and also believing that this was the right career move for me. Originally the financing of this project was linked to the purchase of Nigerian foreign debt; however, during the last eighteen months, the financing has not materialised. We have been kept in the dark regarding progress but have received continuous assurances that all problems have now been solved and the project will get under way in the next few weeks; it never happened. The final straw was the Tony Blair/Bob Geldof debt-relief package for Africa and Nigeria, which has virtually put the cap on this type of debt buy-back scheme.

Although Nigeria is a tough place to live, we could have managed a few more years if the project had materialised and my career continued to develop. If there is one lesson that I have learned when working abroad as an expat, it is to ensure that you work for a large multinational.

I have accepted a post with a British company that builds plants and develops projects in the sugar industry. My first appointment, which will last until June 2006, is for a project in the Komatipoort region in South Africa (for overseas friends about 10 km from the Kruger National Park Gate). Hopefully further appointments in exotic overseas locations will follow. Being employed by a British company, I will be working under the same conditions as a British expat working abroad – that means it will be OK to behave badly, drink too much and complain like hell.

Lagos has been described as being dirty, smelly, noisy, arrogant and ostentatious and although all this is true, it is not a fair description, as its people are generally friendly, and once you get to know and understand the place, you can live a reasonably normal life.

Although Nigeria has large oil reserves and is the world's fifth-biggest oil producer and the people are not starving, in my view, things could be a whole lot better but will not improve until Nigerians change their philosophies of 'live for today', 'might is right' and 'when on top, milk it'. This culture is evident in the lack of any investment in maintenance and the work ethic of doing things just well enough to get by. This attitude has its roots in an intense selfishness, never give or do even a little bit more than you have to, as well as a lack of empathy for their fellow human beings. It is these attitudes that make the ongoing corruption a lot more acceptable to all.

The above can best be illustrated by the example set by the local population in the area where we started developing the site. Instead of seeing this as an opportunity and taking a long-term view and using the money flowing into the local economy for creating jobs, building infrastructure and generally improving their quality of life, everyone from the local headman, chief of Police, residents committee, through to the hooligan youth and church groups have been trying to extort money from either the company or from outsiders coming into the area. They do this through the demanding of compensation payments, informal toll gates and vandalising of site equipment.

I am also of the opinion that many expats are accomplices in the above and are here more for what they can get out of it rather than any real commitment to contribute to the economy.

I grew up in apartheid South Africa, but I have never before seen the level of discrimination and exploitation that I have witnessed here (some expat communities have a lot to answer for).

During the last eighteen months I can attribute the following to Lagos:

- The losing and regaining of my health.

- Experiencing incredible stress due to the uncertainty of the future.

- Stopping smoking and being a lot fitter than I have been in a long time.

- Seeing the worst of mankind, but also man's incredible ability to adapt and survive.

- Making very good friends.

- Inez and I moving closer to one another.

- Coming out tougher and more resilient and knowing that we could probably live anywhere in the world.

In a letter at the beginning of the year, I described our new kitchen and our economical, however tasteless, interior decorating idea of using graffiti to decorate our kitchen wall. Our friends here all assisted, some only after liquid encouragement.

You will all remember that the last letter included a competition, and I have included some of the better answers; however, there is no prize. Embedded in the letter was a code in the phrase 'for one knows that 90% cannot resist a prize', and if you had paid adequate attention when reading Letter Ten, you will remember the description of fraud section 419. Many of you will no doubt be thinking that I have been in Nigeria too long.

I wish to thank those of you who paid their subscriptions, took part in the competition and the many of you who gave me positive feedback and encouraged me in writing my letters.

To all of those who did not come and visit us in Lagos, you are now obliged to come and visit us in Komatipoort, and those who are behind with their subscriptions have to bring the beer.

Cheers and love to you all,

Arthur and Inez

Letter from Nigeria Eighteen

The Last

October 2005

In the last letter from Nigeria, I said that it would be the last, but it was written in a rush, as we were hurriedly packing with a thousand arrangements to make, and as we were still in Nigeria, our Nigerian adventure was not yet complete. As we have been travelling and have been slack on the e-mails, I have decided to do one last short letter to finish off Nigeria, tell you about our European holiday and bring you up to date on our present whereabouts and contact details.

When it became known that we were leaving, we immediately gave our steward notice (she is a very young twenty-one-year-old from a rural district outside of Lagos). Within a few days, her mother came to visit Inez and suggested that her daughter could accompany us, provided we could offer suitable compensation. In my view, selling your child into servitude amounts to slavery, and although it may be that I am repeating myself, that never happened in 'apartheid' South Africa.

This time we decided to airfreight our goods out, and to our relief, all went smoothly. Within a week we had all our goods in Johannesburg. As there is very little airfreight flowing in the direction from Nigeria into South Africa, the cost of airfreight is also very good, less than half that of the importation cost into Nigeria by sea. However, the statistic is skewed, as there was no personal enrichment of the customs officials at the port of entry this time around.

When leaving Lagos, we tried to ensure that all eventualities were accounted for and that no last-minute mishaps could delay our departure and consequently destroy our pending overseas holiday. This included getting to the airport early to make sure that our tickets would get us onto the plane and with plenty of time available to get through customs and immigration. All went well until we got to the last security check. When leaving our apartment for the last time, Inez had grabbed the last two teaspoons, which were ours, and put them in her handbag (regular OK Bazaars/Wal-Mart/Tesco stainless steel teaspoons, not the family silverware). Also in her handbag was a roll of sticky tape we had kept to attach last-minute tags, etc. These were found by the security officials, who wanted to confiscate them.

Inez, determined to the last not to give in to extortion and corruption, questioned this, demanding to know why she could not take her teaspoons on board. She was informed that they could be used to 'hijack the plane'. I am not an operational planner for Al-Qaeda, but at best, the only way that I can imagine an aeroplane being hijacked with a spoon is if the spoon were a wooden spoon and the pilot were my six-year-old nephew. All the while I was standing nearby having a mini nervous breakdown, thinking, all this discussion and negotiation for two teaspoons and a roll of sticky tape!

This reminds me of a friend of ours who was leaving Lagos, and to his dismay, the alarm sounded when he walked through the x-ray machine. He was pulled aside and frisked by a female security guard, who found a gold-plated cigarette lighter in his top pocket. Immediately her eyes lit up as bright as the butane flame from his lighter as she informed him that he could not take a cigarette lighter onto the aeroplane and promptly confiscated it.

Our friend, knowing that the most costly mistake you can make in Lagos is to lose your temper, calmly agreed that he would not take his lighter onto the plane, but he reasoned, I now have a two-hour wait in the departure section of the airport and would like to enjoy a cigarette or two during that time, what does she suggest?

In true Nigerian style, she had a solution. He could buy another Nigerian-made cigarette lighter from her for only US $2.

Having little choice, he bought the lighter from her, never got his own lighter back, fumed through the next two hours, and to add insult to injury, boarded the plane with this new lighter in his top pocket.

This goes to show that when you give people in Nigeria authority, they don't use it to execute their duties but rather to enrich themselves.

After a few days back in South Africa, we were off to Europe. First stop was in England, where we paid a brief visit to my new employer. Thereafter we travelled through the Channel Tunnel, on to Belgium, where we met friends with whom we spent a few very enjoyable days touring Flanders. From there we travelled to Berlin and then on to Holland, where we visited some of Inez's family, and finally back to South Africa via the UK, where we visited old friends. A BIG THANKS to the many people who assisted in making the holiday enjoyable, especially Cecile, Yvonne, David and Dorelle.

Back in South Africa, we travelled to our new home in Komatipoort, which is situated on the far eastern side of South Africa. To the north is the Kruger Park Game Reserve, and on the eastern side, the Mozambique border. The picture below shows our new home.

This picture was taken about 500 m from our house. The fence is that of the Kruger Park, with the Crocodile River in the background. Our morning walks (yes, Inez still drags me around the block) include a bit of game watching.

We have not yet spotted any of the 'big five', but we have seen a plethora of unusual and creepy insects (the area is an entomologist's heaven), a huge variety of birds, a chameleon, as well as the largest snail ever seen by man.

ARTICLE SIX: ARTHUR HAS BEEN HERE TOO LONG

Now we are packing to leave Lagos for ever. Although we have been living in furnished accommodations provided by the company, we took it upon ourselves to purchase an additional table and two chairs to use in our living room. Now that we are leaving, we have given the table and chairs to friends of ours who live about 3 km away.

Although Arthur has been here less than two years, he has already adopted the 'Nigerian style' of moving this furniture. Now I know that you will not believe me, because anyone who knows Arthur would never believe this possible – therefore, I have taken these photographs to help convince you that what I saw really did happen.

Leaving our apartment complex

Arriving at our friend's house

NO ROPES! Arthur holding on to the table through the car window – we have definitely been here too long!

Hopefully some time next year we will be travelling again to foreign exotic places.

Once again, thanks to the many of you who wrote letters and have kept in touch with us during our time in Lagos, it often helped to keep our chins up.

Cheers,

Arthur and Inez